A Dot on the Timeline

Paul V. Mroso

First published in paperback by
Michael Terence Publishing in 2024
www.mtp.agency

Copyright © 2024 Paul V. Mroso

Paul V. Mroso has asserted the right to be identified
as the author of this work in accordance with the
Copyright, Designs and Patents Act 1988

ISBN 9781800949003

All rights reserved. No part of this publication may be reproduced,
stored in a retrieval system, or transmitted,
in any form or by any means, electronic, mechanical,
photocopying, recording or otherwise,
without the prior permission of the publisher

Cover image
© Paul V. Mroso

Cover design (AI)
Michael Terence Publishing

Michael Terence
Publishing

Introduction	1
Growing Up Days	3
In Days Past	49
Family Concerns	89
Worrying Times	139
The Future Assured	181
Reminiscing in Old Age	215
Acknowledgements	244
About Paul V. Mroso	245
Other Books by Paul V. Mroso	246

Introduction

How life systems work and function depends on it
Upkeep of authenticity for generations depends on it
The ability to evolve to survive owes that knack to it
It is the architect of life systems
It is elusive in its functions but obvious in its existence

This elusive entity could have other traits yet unknown. Facial features, body shapes, voices, walk styles and many other characteristics, in human or animals appear to link clan members.

A very old man and his wife had an unhappy thought that their generation was ending. The only hope of passing on the family traditions, character or traits was through a grandson who thought he was their son. However, the many hostile events with a cruel neighbour threatened that hope. A cutting of a thirty-year-old newspaper started a chain of events that eventually ended hostility.
It is a tale of the elusive entity that unearthed traits, through a trail of events to unite a long-lost family.
The old man and his wife were proud to see the increased number of family members sitting at their dinner table. Thanks to the unique traits that made the pleasing outcome possible.
Deoxyribonucleic acid (DNA) is a molecule capable of giving life-enhancing instructions through chemical activity. It is an elusive entity in all live forms. Its discovery brought medical, criminal and archaeological advances. Revered or feared it had the ability to reveal unholy truths.

Growing Up Days

1

You will not be surprised to note that Victor was asking many questions. He was an adolescent growing up, starting to know and curiously wanting to understand things he saw or heard. Many issues became concerns that worried him. Those matters weighed on his mind, changing his mood from a smiling boy to a gloomy youngster. To add to his long list of worries, he was going to a new school that needed the use of two bus rides alone and without his usual school friends. Those fears appear to cause distress. Mama saw the low-spirited face and asked,
"You are going to make many new friends. You like friends, don't you?" Victor smiled a little and asked,
"Do I have to go? I can stay here and carry on farming like you."
"Get your education first. The knowledge will help you to be better at farming than us."
"I want to help Papa because he gets very tired every day after work."
"You will have plenty of time to help after your education. Try your new uniform and see how smart you look."
Victor smiled as he stood in front of a mirror and imagined the admiration he would get among many possible new friends.
During the school break, at mid-terms, Victor was happy to work on the farm with Papa who noticed his zest for work.
Papa said to Mama,
"He works as hard as his father and he is showing signs of maturity. It is time to tell him."
"How will he react?" asked Mama.
"We cannot know until we tell him," said Papa.
"We tell him soon before schools re-open. We will always stay with him and observe him," said Mama.

"He may ask many questions that may cause stress to all of us. We have to tell the bitter truth. He will understand," said Papa.
After dinner, Victor was proud to announce finishing the plot for new crops. Mama patted him saying,
"You always do good work. Do not go to sleep yet, I have something to tell you."
Victor's eyes opened wide and waited expecting something like a reward for his work.
"Have you ever thought why we look older than the parents of your friends?"
"No, Mama, I do not think you are old at all," said Victor.
Papa said,
"What about me?"
"Not much," said Victor still curious.
"Did you not think we could be grandparents?"
"No, I did not. If you are grandparents, where are my parents?"
Papa and Mama were silent. Victor asked again,
"Please tell me if they abandoned me or are they in some kind of trouble?" Papa faced Victor, touched both his shoulders and said,
"Your father was my son; your mother was from another area in a town, far away from here. When you were three, they were travelling to town for business but a motor accident took the lives of both away. We took care of you since. We think now is a good time to tell you because we think you will understand."
Victor sat silently probably stunned. He must have started to imagine the features of his parents. He did not cry. He had his mind weighed by many thoughts of questions that needed answers. He stood and started to walk away as if he was going to sleep but returned and asked,
"Why did you keep that a secret for ten years? Mama touched his hand, pulled him closer and said,
"You were a child; we thought you could not understand such harsh information. Now we can answer all questions you may

have." Victor appeared to understand Mama's point of view as Mama embraced him.
He started to cry.
Papa and Mama held Victor as he cried while with teary eyes, they thought about the many questions he would eventually ask. His cry put an indelible mark on his mind and a spot on his time of family life. He was probably not aware of how that information could influence his life. Many questions passed through his mind as he cried. How would life be, was a question that probably dominated the young mind. Mama's embrace loosened as the trembling body of Victor indicated that he was less agitated. When Mama and Papa looked at Victor's face, they detected a changed person from a child to a young man.

It was inevitable that Victor would change.
The knowing was a dot on his timeline
The info could shape his future

"My other grandmother can tell me about my mother. The information may help me to picture her in my mind. Have you any pictures?" asked Victor. Mama looked at Victor with her teary eyes and said,
"In those days it was too expensive to take a picture and we could not afford it."
"Can I meet my maternal grandmother?"
Papa faced Victor and touched his chin saying,
"We tried after the accident. They had moved from the address we knew. We made enquiries but we never heard from them. They would know about the accident. That serious accident raised many questions of safety and was in the papers for a long time."
"What are their names?"
"Vince and Josephine were their names. I do not know why they did not make contact."

"Did she have brothers or sisters?"
"Vince and Josephine came to the wedding. They were alone and there was no mention of brothers or sisters. Vince and Josephine talked very little about themselves and we could not ask too many questions especially when we had to help them to pay for their bus ticket to go home. Your mother Charlene did not talk about brothers or sisters. She was the only child of Vince and Josephine."
Victor scratched his hair. He was thinking about how there was no past he could find to link with the family of his mother.
"Papa and Mama, I hope you will be able to tell me a lot about my father."
"Yes, there is a lot to tell but that will be another day," said Mama.
"Will you take me to see the eternal resting place for my father and mother?"
"We will go tomorrow," said Papa. "You can now go to sleep. You know we will always love you," said Mama.
Victor stood up and said, "I know," and left to go to try to sleep. You would guess wrong if you assumed Victor had a good sleep. It was anything but that. He twisted and turned. He asked questions he would ask Papa and Mama but on more thought, he discovered he had answers. He grimaced at thoughts of motor accidents, imagined from piles of mangled trucks he saw on roadsides as he travelled to school. He wanted to know where exactly the accident took place. He reserved that question for the morning. The bright sunshine outside reminded him of the talk the previous day. He started to tremble and to sweat. He remembered it was a weekend. He was not travelling to school. It was a mid-term break but instead of relaxing, he became more agitated. He remembered what he asked Papa and Mama to do that morning. His sweat increased. He threw away his bedclothes, went at speed to the shower and returned whistling

to try to uplift his subdued mood. He dressed at speed. He went to wake up Papa and said,
"Papa you promised to take me someplace this morning." Papa started to prepare to keep the promise. He finished dressing and was ready to go when he found Mama with Victor drinking a mug of coffee and chatting hilariously. He thought that the joyous mood would help to lessen the trauma of what they as a family were soon to encounter. Victor said,
"Papa you are not going out with your farm boots. People will laugh at us." Mama did not say a word; she knew why Papa was not formally dressed. Papa thought of a mood-raising witty chat. He patted Victor's shoulder and said,
"We will use shank's pony."
"What is that mama?"
She laughed and started walking. Papa walked past to the front as he said to Victor,
"We walk."
Victor walked slowly remaining at the rear, as they walked in a single file following a footpath that resembled the winding as a snake sand path. They walked through the woodland part of their farm that was unfamiliar to Victor. Papa had left that part of the farm to grow naturally.
They arrived at a mound that was fenced by trees. They stopped; Papa turned round to face Mama and Victor. In a solemn tone of voice, he said,
"This mound is a spot where four farms meet. We agreed to set an area for our departed."
He turned and walked forward into the enclosure. Victor saw at a distance only two graves but quite apart from each other and asked,
"Why are my parents buried so far apart from each other?"
Mama held Victor's hand and walked to one gravestone and stood there saying nothing. Victor read the stone:

A Dot on the Timeline

"Danny and Charlene": the rest of the reading was very clear to him. When Victor asked why his parents were together in one grave, he was shocked as Papa's tears started to fall on the gravestone. Mama said,
"Let us go." then he whispered to Victor and said,
"I will explain to you later. It is too traumatic to talk here."
Papa placed on the gravestone the single flower he always took on visits, said some words and left. He walked very fast leaving Mama and Victor behind. Mama said to Victor,
"You can now see why we did not tell you. The horror of the accident is still raw in his mind. I hope you will overcome all you know now and carry on."
"I will try Mama," said Victor as they arrived to see Papa with his tools going to the farm to check and remove the stubborn weeds.

2

The last year in school was the time to work hard educationally as Mama encouraged Victor to aspire to higher learning. It was a difficult time for Papa and Mama. They were living alone in the farmhouse. They did not want to stop Victor from attaining his potential. They did not know how life would be without Victor around the house if he left to go away to the city to work in an office. They had to make that sacrifice. Victor had to get his education. The use of student accommodation was necessary, as the college was too far to commute daily. During holidays, Victor was encouraged to work in the firms that enhanced his studies. That meant for three years, Victor met his grandparents once or twice a year. The loneliness had made them weak. The farm started to fall into a spiral of deterioration. Pressure from aggressive buyers made Papa and Mama despondent. It was only the memory of how they acquired the farm and how hard their departed son worked on it that gave them some energy to resist selling. Some neighbours who supported Papa to resist selling the farm were an asset. They served as a prop that kept one standing.
Mama woke up one morning. She was as active as a lamb that had just a fill of suckling. She was fully dressed as a farmer's wife would. She had an idea that sparked action with zest. She did not remember what she ate the night before but she was energetic. She shook Papa; he was still asleep and said,
"I had a dream. I was going to die in this bed if I did not wake up and go out immediately to do something to show I am alive."
Papa granted,
"You and your dreams, please let me sleep."
Mama shook Papa again and said calmly without raising her voice,

A Dot on the Timeline

"I know we are down mentally and weak physically. We have no help. Are you going to sit down waiting to die or do something to show we tried our best before we die?" Papa rolled slowly out of bed. He appeared to understand and agree with his wife. He did not sulk or moan. He thought of Victor and said to Mama, "We should not allow Victor to find the farm neglected. He is graduating soon. Let us show him that we are working as hard as he did in college. Let us clear the whole farm and plant." Mama smiled assured she had started the ball rolling. Friendly neighbours saw the old couple struggling and offered support. The transformation of the farm surprised many and made those with the intention of buying Papa's farm very insecure. The confidence of Papa and Mama was high but they did not want to gloat when Victor arrived. The site of silos full of grain, the stacks of hay that filled the barns and the grapes waiting for a buyer, were what Victor saw when he came home as a graduate.

3

The smile could dazzle anyone looking. The arrival of the students was in droves not in trickles. It was graduation day. Everyone dressed perfectly. The ceremony and the merriments that followed made an indelible mark on their path of life including that of Victor. The formal occasion, prestigious and pleasing was a milestone in Victor's timeline of life. Victor stood alone holding his certificate rolled in a cylindrical hard cardboard. He was waiting for his chance to take a picture dressed in his graduation gown. He was expecting his sweetheart, Titti to join him. The other graduates, accompanied by parents were all in smiles as they chatted. In their pose for the photos, Victor observed the sheer bliss of his colleagues flanked by one or two parents, relatives or sweethearts. Victor shed a tear, he was alone and he had no parents to hug, talk to or to congratulate him. He was about to leave and go elsewhere to avoid the smiles of parent sibling cuddles, when someone tapped his shoulder. He turned to see Titti. She saw the teardrop and hugged him.
"Why are you crying?" she whispered.
"I was happy to see you," said Victor knowing it was not the whole truth.
It was after the photo taking that all the graduates including Victor threw their graduation caps up into the air. Victor caught it as it tumbled down and he thought it was a good omen. It did not matter but that was what he thought.

Hand in hand, Titti and Victor planned to leave, hurriedly saying goodbye and forsaking all the invitations to parties or meal sessions. He was keen to show his graduate certificate and have dinner with his grandparents. His eagerness to reach home was what made that short journey appear very long.

He did not forget about his teardrop as he sat in that coach alone after parting with Titti. The coach appeared to move very slowly as Victor started reflecting with sadness the events that just happened.

> *A life-changing incident is indelible*
> *The tear shedding was from such a memo*
> *That occurred over two decades ago*

Victor approached Papa and Mama smiling but their gloomy faces made him pause. Thinking Papa was unwell; Victor approached Mama and asked,
"Do you think he would enjoy a cup of tea?"
"If he refuses then he must be very ill," said Mama smiling as she was encouraging Victor to make the tea. Papa smiled as Victor placed a cup of tea on the stool next to his chair. The offer or the cup of tea appeared to make Papa cheerful. It was like a tonic. In an elated mood, Papa said,
"You must have passed, you look very happy."
"Yes Papa I passed with honours," said Victor and without wasting time, he unwound the rolled certificate and stretched it in front of Papa to see and then he talked hilariously about his achievement. Papa and Mama listened as Victor outlined a future with a job, a good income and a chance to build his own house. To Papa and Mama, Victor's plans increased their worries. The plans showed little enthusiasm for farming. That thought made Papa and Mama fear for the future of the farm as Victor's plans did not include any hint of farming or remaining in that farming community.

Victor noted the subdued response to his talk. He did not know the reason. He changed the subject saying,
"Papa and Mama, I have made some friends. I have a special one, a sweetheart. Can I bring her here to meet you?"

Papa smiled with pride and held Victor's palms tightly. Papa's grip was weak and his hands were cold. Victor was scared that Papa could be ill. Mama's warm and passionate embrace was pleasing as the three family members chatted hilariously until dusk. Mama stood up intending to go to her room but she said to Victor,
"When shall I see this sweetheart of yours?
"Soon, Mama you will see her. Tonight we are going to celebrate my graduation and we will talk about the day she will come for a visit," said Victor smiling.

4

The farmhouse built on a mound was visible from distant places and the veranda faced the farm that extended to the horizon as far as the eye could see. For the last three years, the farm was in a gradual decline and in desperate need of restoration. When Mama and Papa with the help of neighbours and friends made an effort to restore it to please Victor, Papa was sure it was a short success as Papa's energy was failing with age. That was probably the cause of Papa's gloomy face. Papa and Mama were sitting on the veranda sipping cups of tea. Papa looked at Victor, pointed at the horizon and then said,
"One day all that will be yours. You will make it a farm to be admired by many and may provoke envy to some."
Victor was barely a mature man capable of taking great responsibilities. He sat next to his grandfather silently but full of worrying thoughts. Learning and managing a farming business was challenging. His dream of a comfortable city job could end. Would his sweetheart agree to carry on farming? These issues took over Victor's mind as he sat next to his grandfather looking at the immense farmland. He failed to comment on his grandad's suggestion about managing the farm. His grandfather probably understood it was a shock to the young mind and he did not insist on an answer.
Victor had lost both his parents in a tragic motor accident as a baby. The grandparents whom he always addressed with respect as Papa and Mama took up the parenting duties up to his adulthood. The community considered Victor an adult. Papa and Mama knew that he was not mature enough to take on the big responsibilities of managing a farm but he was the only fresh energy source they depended upon to revamp the farm. He was however mature enough to have a girlfriend and that was

comforting when he introduced her to his grandparents. Titti, his sweetheart, became a new source of energy and happiness for Papa and Mama. Papa said to Mama in a whisper,
"Are you seeing what I am thinking?"
Mama smiled and said,
"I can read you like a book. You are thinking about the story from a book you just glanced at the title but I am reading the story."
"What did you see?" asked Papa teasing Mama.
Mama smiled again and said,
"Lend me your ear. I see many great-grandchildren running around like schoolchildren in the playground."
They both burst out in a peal of hilarious laughter that alerted Titti who asked Victor,
"Why are your grandparents laughing?"
"Are you happy Mama?" asked Victor.
"Tell her she is very beautiful my dear," said Mama in reply.
"She is a beauty, Mama, I know she is," said Victor happy that the adults showed great pleasure.
Mama asked,
"How did you two meet?" Victor had to think fast. It was over six months in the past. He whispered and asked Titti,
"Do you remember how we met?
Mama heard Victor's question and said, "Men have a bad memory. You tell me, my child." Titti smiled and said,
"We met inside the bus when a naughty boy tripped me and I fell. I saw a hand stretched for me to hold to stand. He gave me a white handkerchief to wipe the dust. When I returned it to him, he told me to keep it. I knew he was a kind man and I told him my name when he asked. We got on well until now I have the chance of meeting you."
Happiness can do wonders for people. The thought of a wedding was one of those wonders. The family progression was another. The thought that one day Mama could hear the cry of a baby in

her house was what made her smile. She wished the young couple to be interested in the farming business and stay, rejecting going to the city for a job.
She thought that possible love for farming would keep them at that home and in the big farmhouse. Papa was thinking of the profit and satisfaction in producing one's food compared to paid jobs in big cities where expenses usually outranked income. He wished that Victor would see his point of view. Victor, however, was in deep thought about his future. He started by questioning the purpose of his high learning.

With a credential of high learning
That offers a choice to seek jobs or to be a boss
Moreover, that needs a deep thought
To decide now for the future

Victor and Titti decided to stay in the farmhouse as they waited for job offers from their numerous applications. The more they stayed the more Titti loved Mama for the care she showed towards her. Victor got more and more agitated at the failure to get a job. The fear of losing Titti if he did not show more affection was another issue. Mama saw Victor's tension and gave a word of advice,
"Victor, my grandson, I know you are tense, waiting for job responses. Do a bit of farming. You may like it because it was your father's farm. I am getting fond of Titti. I wish she could stay forever."
Early in the morning, Victor went to talk to Mama. Papa heard a burst of loud laughter from Mama and then an ululation a sign of great pleasure. It did not take long when Papa heard,
"Papa, come here quick. The two have agreed to get married."
Papa grabbed and embraced Victor. He rarely did that but Victor wondered where the old man got all that energy. Victor was very happy when he heard,

"You have made us young again. I will have the strength to work on the farm as before."
Mama added,
"We sincerely hope you will help us."
Victor and Titti were on cloud nine. They were in love and barely heard what Papa and Mama said.
They were aware of the constant giggling of Papa and Mama and at times an ululation from Mama, a sign to announce her great pleasure.
Papa planned a small but lavish wedding for his grandson and Titti. The tender care, the love and the gifts given to Titti were probably some of those inspiring factors that made Titti wish to be a farmer. She shared those thoughts with Victor who showed his pleasure by smiling but not in full agreement.
The wedding was for close friends and neighbours only. Titti had a small family, hence the small gathering. All who attended the wedding of Victor and Titti witnessed how the wedding had changed the frail grandparents to becoming active and exciting. It was the few times when Victor witnessed his grandparents dancing. In that state of excitement, Titti stood to accompany Mama dancing. There was an expression of love. Papa and Victor were very happy to see the two women talking in such a warm and friendly mood. In that state of excitement, Mama called Victor to approach and said,
"You will look after this woman and love her as you promised. I will not let you make her sad." Titti smiled with happiness at the support she got from Mama.

5

It is an astonishing thing, when you are enjoying your success someone could be planning your fall. Could that happen at the wedding of Victor and Titti? Could you imagine how an unsocial incident could affect the happiness of Papa and Mama?
Papa had invited a man called Bosco not because of his pleasantries but purely out of courtesy as a neighbour. He was a neighbour hated by Papa and Mama including many others in the farming community. He was rich, conceited, cunning but not very clever. His unsocial acts were a constant threat to all the neighbours. Papa thought that snubbing him could further spoil the fragile peace shown when he heard of a wedding. He wanted to get an invite to the wedding. While Bosco sat and enjoyed drinking, Papa could imagine what Bosco was thinking about. Papa thought that Bosco was secretly rubbing his hands in glee in anticipation of some gains. Papa's farm was a hot property that Bosco aimed to acquire. Bosco was thinking that the youth and his wife would soon go away to the city to take up jobs after graduation making it easy for him to buy Papa's farm. Victor knew Bosco as a bad neighbour but he reserved his opinion for later. He was too busy to care at that moment of his wedding. When Papa glanced at Bosco and saw the smile or rather a grin, he imagined that Bosco was planning something evil. He could not be further from the truth. Continuous deterioration of the farm could make it easy for a takeover. That thought worried Papa as he could read Bosco's thoughts from his body language. The rude man could be thinking about the farm selling cheaply as the frail couple, unable to do the hard farm work, would have no choice but to sell.
While Bosco was in thought, probably how much he would offer, some keen eyes were watching him. It is common in social

encounters like wedding parties; people glanced at one another to check the appearance of happiness, sadness or the attire. Goodwood, a friend and a neighbour of Papa and Mama, saw Bosco and recognised him as a childhood schoolmate and a playground bully at junior school. He remembered his gloomy face and was surprised to see him with an occasional smile. He remembered that Bosco smiled only when he was thinking or planning something evil. He planned to ask Papa about Bosco after the celebrations. Bosco was cunningly checking whether anyone was looking at him. He noted Goodwood exchanging glances with Papa but because he did not recognise him, he did not care. Goodwood however, was determined not to allow Bosco, the school playground bully, the freedom to practice his unsocial habit to the neighbours of the farming community and especially to the vulnerable people like Papa and Mama. Goodwood kept his ideas to reveal to Papa later.

Papa was in a good mood after the wedding as that was all well. Victor had not received any feedback from his job applications. Papa thought of introducing Victor and Titti to the world of farming that involved muck and not-so-clean nails. He hoped that they could like it and stay in the farming community. It was a gamble. Would Papa succeed?

Victor and Titti woke up. It was early in the morning. They heard Papa's call. Dressed in boots and overcoats, accompanied by Mama, Victor and Titti were ready to take a guided tour of the farm. After a long walk, Victor saw grazing sheep on his land and asked,

"Papa, why are the neighbour's sheep in our land?"

Papa gritted his teeth. Victor thought it was not going to be good. He asked Mama who said,

"The sheep belong to Bosco. He is not a friend but the worst of the neighbours who want to buy our farm. Those acts are meant to push us to despair and eventually sell the farm to him."

Victor asked,

"How much was he prepared to pay?"
Papa and Mama did not reply. They ignored Victor's question, as they were not happy. Victor had thought naively that it was a brilliant idea. He simply thought that the sale could discharge him from farming. He could move to a glamorous clean city job. That idea quickly stopped when Victor heard Titti saying,
"I love this farm, Papa.
Please do not sell it. It is better to work here than in the city where we work with the constant threat of dismissal for lateness or other misdemeanours. We will stop the illegal acts against the farm, be it from Bosco or any other neighbour."
Victor quickly abandoned those juvenile thoughts when he found out that Titti saw potential in farming. Victor had to back down and show his grandfather's enthusiasm towards farming by repeating what Titti had said. He had to behave like a boss. That idea was frightening but gave him a sense of direction, purpose and pride.

6

Victor and Titti spoke devotedly to make the farm attain its great potential again. That talk was a source of happiness for the grandparents who walked as if they had springs in their steps. When Papa was happy, Victor thought, it was a likely time to celebrate. He whispered to Titti, leaving her giggling as he sped to the house ahead of his grandparents. A bottle of wine in a chiller was ready for Papa as he arrived at the house.

The neighbours including Bosco and Goodwood had observed the farm tour.

Papa, Mama, Titti and Victor were at their veranda discussing farm issues when Goodwood arrived and said,

"I saw Bosco's sheep on your farm. He inflicts unpleasantness on all neighbours. Please give some advice on how to stop him."

"Have you got something against him that you could take legal action?" asked Papa. Goodwood was an academic, who spoke clearly, in a polite language unlike most farmers and said,

"I was having a barbeque with my colleagues from the city when Bosco sprayed water forcing all to go inside."

"What was the reason?" asked Papa.

"He did not like my friends because it was those kinds of friends from the town that made him penniless and created a life of misery forcing him to abandon the town."

"Did he know them?" asked Papa.

"He assumed anyone smartly dressed was a city resident and he hated them," said Goodwood.

"That is how a bully would think and act, as anything superior to him becomes a threat," said Papa.

Goodwood replied saying,

"He was a bully at school. He thinks that it is as easy to bully someone in adult life as it was in childhood."

A Dot on the Timeline

"We can accuse him of spraying you with water," said Papa.
"It will not help because he used the spray that went high up mimicking raindrops. Most people thought it was rain," said Goodwood in desperation.
"We will find a watertight proof to accuse him," said Papa.
"Did you have a feud before the attack?" asked Victor.
"Yes, I had refused to sell my land. You, Papa, told us to resist his approach," said Goodwood.
"We will curtail his unsocial acts," said Victor.
Goodwood left happy that he got support but the visit make Victor determined to protect his grandfather. Papa would be alone and could fall victim to Bosco's intentions. He could not allow Bosco to force Papa or other neighbours to sell.
A season of uncertainty passed as Papa managed to plant and harvest as before with Victor carrying out farming with his grandad:

It was a bold decision
Happy despite the failure to get a city job
Proud as a farmer and a boss
A change in thought that was pleasing

The family ate the evening meal in glee and Titti made the situation more hilarious when she said to Mama in a whisper,
"I am expecting."
Mama stood up and said,
"Victor, go to the cellar and bring a bottle of sparkling wine."
Victor obliged but Papa complained saying,
"I do not want more wine; I am having a beer tonight." Mama looked at him, smiled and said,
"You will not taste a beer tonight. You will have three glasses of the wine when I tell you the reason."
Papa was silent observing and following the sequence of events keenly. Titti watched as Papa was surprised and smiled. Victor

brought the wine, wiped the bottle, opened it and poured in three glasses. A small cup of orange juice was for Titti. Mama gave Papa a glass and said,

"Drink to the health of your great-grandchild." Papa looked at Titti, smiled and gulped the wine. Good news makes life respectable. Papa started walking around the house whistling his memory lane song and then sat on his rocking chair and said,

"Victor and my dear Titti, you have made an old man very happy."

It was probably the wine, the relaxed mood or the jumping and whistling around the house that made Papa fall asleep as he sat on his rocking chair. That brief moment allowed Mama and Titti to talk before they went to bed.

Papa said to Mama,

"What are your plans?" Mama sat up from a reclining position on the bed and said,

"I know what to do. It is you who has no idea of what to do."

Papa was not happy with Mama's comments and said,

"Frankly, you girls think you know all and do all things better than men. I will carry on farming until the child is born. I will know a lot of what to do by that time."

Mama detected a temper and said,

"Carry on farming to ensure enough food for your growing family and I will do the pampering of the mother and that will keep me busy."

"Go on do your best and let me sleep," said Papa as he turned away from Mama and pulled the bed covering to sleep.

7

The birth of Alena, a great-granddaughter was as if the heavens had showered that household with a fortune of finding gold on their farm. The whole family focused on the new birth. Papa and Mama started to think of the future, something they had not done for a long time. They felt rejuvenated and happy as if they receded to their youth like Victor and Titti. Papa and Victor constantly meet to talk about family and values. Victor was no longer a child as viewed by Papa and Mama but an adult whose ideas gained respect.

Many activities of the family focused on Alena. They did not have time to confront Bosco who continued to cause unpleasantness. The family kept happy by thinking of the good news. The baby, the mother and their welfare weighed as essential. Papa, Mama and Victor noted Bosco's tantrums but they did not respond. The lack of any action from Papa made Bosco nervous, as he would think that Papa would sell to another person. He had to know what was happening in Papa's household as no one paid any attention to his abuses. He approached Papa, who was walking and whistling. The happiness Papa showed increased Bosco's agitation. He had to know but was afraid of asking a direct question. After a brief moment of hesitation, he approached Papa and asked,

"Are you well? I do not see you outside often." Papa knew that the question was not out of concern for his welfare but Bosco had something sinister in his mind. He paused and then said to Bosco,

"I am not dead yet. I am a recharged old man. Get a mirror, look at your face; you almost look as old as me."

Bosco stood still. He probably did not understand the significance of what Papa said. The insult about his age, however, made him walk off in a huff. Papa's good mood was at a fever pitch. He walked home slowly, whistling as loud as he could for Bosco to hear. At home, he was proud to tell his family about his encounter with Bosco.

8

There is pride, self-assurance and at times arrogance after gaining a high scholarly qualification. Victor was not immune to many such temptations. He was aware of many choices open for him after graduation. The opportunities were many and at times, they were confusing putting him in a complex dilemma. He knew very little farming. It was not what he learned at college or what he wanted to do. However, he was sure that Papa and Mama wanted him to stay at the house and carry on farming. He had imagined good high-paid salaries in city jobs. He thought of buying a car. The social life in the city kept him in thought as he stretched on the lawn trying to make a decision. Papa and Mama were sitting on the veranda talking in low voices. They had not seen Victor who heard as they spoke.
Papa said to Mama,
"I will admit to you that I am worried."
"What can a strong man like you fear?" Mama asked.
"If Victor goes to the city, it is the end of farming for the family. I cannot bear the thought of such an end," said Papa.
From the corner of his eye, Victor saw Mama moving closer to Papa. I think they are sad, he thought. Then he heard Mama saying,
"We cultivate a small area to provide us with enough food to survive."
Papa was a little louder. It might be from fury as he said,
"When people see the deteriorating farmland, the pressure to sell cheap will increase and I will have to sell the farm that his father put great energy and sweat to make it sustainable. That thought will kill me."
"When we die, squatters could move into the uncultivated land and make camps, it could be lost forever," said Mama.

Papa said,

"The thought of that happening will send me to the grave sooner."

"We should ask Victor for his opinion. He will help us make the final decision," said Mama.

There were no more exchanges and the silence that followed made Victor uneasy. He realised that his heart was racing.

The sun was setting and it was getting cooler, forcing Victor to go inside to talk to Titti. They were close to talking about what Victor heard but they heard Papa calling.

"We are coming Papa," said Titti.

Victor and Titti sat next to Papa and Mama. They were surprised to hear Papa say,

"Do you want a cup of tea?"

The surprise was that they did not know if Papa could make a cup of tea. Instead, Titti offered to make it. After a couple of sips, Papa moved to ensure he was at eye level with Victor and then said,

"You and Titti will carry on and control the farming activities from now."

Titti stood up, looked at Victor, nodded to signal acceptance and said,

"We will try our best Papa." Mama stood up and embraced Titti. Victor's legs were wobbly, but he stood up. The troubling thoughts of responsibility had come to reality. So far, he had worked with Papa. He did not have to make plans or decisions. He tried to show a bold attitude. He kissed Mama, shook Papa's hand and said,

"I will do it Papa, but please be ready to give me lots of guidance."

Papa stood up and locked arms shoulder to shoulder with Victor. On their way to the office to look at the paperwork, Papa's fear had gone and with confidence, he said to Victor,

"Do not worry; the work is as easy and as interesting as writing a letter to your sweetheart. You will make the farm a shining example again."
"Papa, you make the work sound amorous. Is that how you feel?"
Papa chuckled saying,
"There is nothing as exciting as following the process of plant growth from seed to harvesting." He then whispered to Victor saying,

> *You sleep and your wealth grows*
> *You gain riches as it rains or shines*
> *You are truly the boss that nature guides*
> *How can you not love that* state of affairs?

"I did not look at farming with that pleasing thought," said Victor.

Papa went on to advice saying,

> *In office work you sleep, the job waits*
> *Year after year, you sit at the same unchanged desk*
> *The four walls of the office unchanged*
> *No wonder some get mad*
> *But*
> *In farming, however, each day is unique*
> *What you see has increased in size*
> *Give a new colour or develop a new shape*
> *That generates excitement or sheer bliss each day*

Victor listened and eventually said,
"I will make you a contented man by doing it well, Papa."
Papa said to Victor as he slammed the folders on the dusty table,

"You will study these records. They are for you to read tomorrow. For now, I have more to show you." Victor noted that Papa was moving fast doing things efficiently without hesitation and he thought,

"He is so happy; I will not do anything silly to end that happiness. I will discard the thoughts of city jobs."

When Papa pushed to move a stand-alone picture frame, it revealed a small dusty concealed door. On opening it, the screeching startled Victor. He moved a step backwards worried it was something dangerous.

Papa turned to face Victor and asked,

"Have you been down here? I would like you to get accustomed to many noises in a farm such as the hissing of insects that live in dark corners in basements, unlubricated machinery, hinges or the feel of cobweb touch."

As Papa switched the lights on, Victor saw many bottles, jars, pots and sacks with contents covered by cobwebs. There were rusty machine parts, farm tools and numerous containers of oils and cans of lubricating grease. He looked at his granddad and said,

"I will come in the morning to clean the basement." Papa smiled as he turned to walk to the far corner to make sure Victor did not see the smile of thrill.

"You must know where everything is kept. You must keep some seeds in case of crop failure and you must store enough food for your family," said Papa as he turned to look at Victor.

Victor and Papa, after a long talk of do's and don'ts, returned to find Mama and Titti having a hilarious conversation.

Titti in excitement said to Victor,

"Mama and I with the baby went to the attic for a tour. Mama showed me nicely wrapped bundles of clothes and other valuable goods. Mama said that Papa saved his wine and seeds in the basement."

"Papa and I were in the basement and this is one of the wines," said Victor as he wiped the dusty bottle. Titti interrupted saying, "Mama said they used to make wine years ago before you were born. She stopped due to the hostile activities of Bosco. Mama found it more profitable to sell the grapes than to make wine."
"In business, you change to optimize the chances of success," said Papa.
Victor laughed and said,
"Why did you give in to pressure from that malicious dimwit?"
Papa and Mama looked at him in silence and Victor realised he had said something improper. He started serving the wine to avoid that awkward moment. They talked about the taste of the wine, a subject that pleased Mama and Titti.
Victor looked at Titti and said,
"Would you like to learn that skill of winemaking?"
"We have started talking about that already," said Titti smiling at Mama.
The soothing light of the full moon or the vintage wine probably made Papa doze off in his rocking chair. Looking at him and glancing at Mama, Victor thought.

They are passing to us their experience
For us to experiment and practice
To build a future for our preference

9

Victor went to bed but could not sleep. Would you fall asleep with so many heavy responsibilities on your mind? He got a slight jump of his heart as he recalled the acceptance to take farming duties. He had to give up, half-heartedly, his plans for a smart, glamorous city job. The process of decision-making was easier as Victor noted that his wife and child were in a secure home. He decided to give farming a chance. He listed the assets on the farm that he could further exploit to make the farm profitable. He relaxed and was about to fall asleep when the name Bosco sprang into his mind. The thought of shooting him dead would not get acceptance from Papa or anyone. The idea of turning him into a friend made Victor cringe as a person with a phobia of the sight of spiders. In that state of recoiling, the baby cried in need of a feed. Titti woke up to find Victor sitting on the side of the bed. She asked,
"What is the matter?"
"The baby was crying, I was about to pick her up from her cot to bring to you," said Victor, knowing it was not the whole truth.
"Go and get her," said Titti.
"No, I cannot because I am trembling too much."
"Why?" asked Titti.
She went to pick up Alena from her cradle waiting for a response from Victor. The baby finished feeding and was back to sleep when Titti asked again,
"Are you worried? What is making you tremble?"
"It is Bosco. He has given Papa grief for many years. It has to stop," said Victor.
"How are you going to stop him?" asked Titti.

"I have no plans yet. I am afraid of failing to control or stop his actions against the farm. It appears Papa is too nonaggressive," said Victor.
"We have one friend, Goodwood, together we will stop him," said Titti. Victor rubbed his hair and said to Titti in a humble worried tone of voice,
"Do you think we can succeed in this farming venture?"
"Mama told me that she will work hard to ensure we prosper in this venture," said Titti.
"Papa assured me that he will be there to help," said Victor. Titti realised that Victor was in doubt of his capabilities. She held Victor's hands and said,
"These hands have written examinations and passed, they have made tea for grandparents and they enjoyed, they held my hands and I followed. What kind of success do you want to see? These hands are a success. Go to bed, get a night of good sleep, wake up in the morning and start farming."
Victor regained his ego and in a murmur said,
"It is good to have a wife like you. I will use my hands and my mind to conquer all hurdles." Before she switched off the light, Titti glimpsed and saw Victor smiling. She said nothing but was sure he got the message and then she said,
"Did you know?"
"Know what?" retorted Victor.
"I am planning to plant more vines as Mama and I will start to brew wine again," said Titti smiling.
"That will make Bosco go as hot as a scotch bonnet in temper. He may go mad, sell the farm cheap in anger or become violent and we may have to take him away to an asylum. It will be good riddance," said Victor.
"I know, it will be a clear message that we mean business and you can tell him we plan to buy his farm and that will frighten him," said Titti.

"We will fight at a position of strength," said Victor chuffed by the idea. After a thought, Victor said to Titti,
"We tell Papa our idea."
Titti replied saying,
"We should always tell Papa and Mama to ensure we act as a bonded unit."
The conversation between Victor and Titti raised their vigour not only to farm but also to expand the cultivated area of the farm and to increase the variety of crops to cultivate. In the excitement, they did not notice that the room was bright or that the birds were singing very loudly. An occasional bleating of goats and the moo sounds of the neighbour's herd of cows alerted Victor to get out of bed abruptly. It was dawn. He had tasks to do. The distant sound of a tractor was another alert. It was time to prepare the farm for planting. He took a pen and a small notebook to write questions to ask Papa. A cough was a sign that Papa was awake. The wearing of farm boots by Titti surprised Victor. He heard Mama's laughter and that added to his morning surprises. He dressed fast and went out ready to confront whatever the farm throws at him.

10

Victor started to behave like a man in a competition, struggling to become a winner. He had rejected the idea of getting city jobs and had chosen affirmatively to become a farmer. He looked up to his grandfather and wished to gain his respect as his father did. The speed at which Victor got out of bed, which was unusual, confirmed his state of mind that he wanted to win. Titti noticed the rare behaviour and asked,
"You are full of beans. What is the secret?"
Victor smiled and said,
"Would you believe it if I tell you that I stopped taking the foo-foo pills from last night?"
"I do not believe there is such a thing but good luck, they were not good for you anyway, I like what I see now," said Titti.
"I am going out to check the farm and plan a future," replied Victor walking away. He stopped to hear Titti saying,
"I am going to check the vines,"
"Ok, see you later," he replied rubbing his hair pleased, that Titti was excited too.
Papa and Victor strolled to the farm. Victor impressed his grandfather with his ideas on development. Papa was taking a back seat in the farm management as Victor heard,
"You are the boss now; I will give guidance as you request."
Titti and Mama burst into laughter when they saw Papa and Victor approaching. Their walk showed that they were happy. Their smiles exposed their teeth as people advertising toothpaste, suggesting a mutual understanding. At earshot, Mama could not hold back her excitement as she asked,
"Are you in agreement with all your farming plans?"
"Calm down woman, it is only the first day," said Papa.
Mama commented,

"Victor, you now look professional as a farmer. You will make a good boss."

Papa and Victor, Mama and Titti were walking to go home. They were exchanging information about their farming plans and then Victor suddenly walked fast to the front, turned and said,

"I have to start the ball rolling." Victor's enthusiasm delighted Papa and Mama while Titti was very proud of all she heard. Victor's focus was on the farm. The failure to get a city job had probably changed to become a reward gained by making Papa and Mama cheerful.

The sounds of machinery showing activities in many areas of the farm pleased Papa and Mama. Victor noted the curiosity of the neighbours and said to Papa,

"Why are they snooping and asking many questions?"

Papa replied,

"Be careful, it is admiration to some but others with a hint of envy are planning to see you fail."

The energy and vigour that Victor put into his work pleased Papa and Mama who saw their planned retirement progressing successfully and gracefully. Alena was a source of happiness as she was healthy and growing well. It was exciting when she progressed to walking and talking. Mama and Alena spent many times together while Titti was busy brewing.

Victor's confidence was at fever pitch. The high yield of produce harvested made him feel proud and confident. He called himself a farmer. The overflowing silos, full of grain were proof to Papa and Mama that their grandson had achieved the desired goal. Titti was happy to claim the title of a brewer as the first buyer returned for more proclaiming the high quality of the wine. Bosco was green with envy.

The arrival of several large trucks to collect produce amazed many neighbours who watched, probably with envy, the loading that took a while. A Truck labelled WINES in red letters attracted Bosco's attention. It took longer to load and when they left,

Bosco saw Mama embracing Titti to congratulate her. Papa and Victor decided to go for a walk around their farm to calm down from the excitement of a good sale. Bosco noticed all the goings-on and could not control his envy. He approached Papa and said, "I see your buyers have come early. Did you give them an incentive?"

Papa was grinding his teeth and Victor was boiling with temper thinking about how to respond. Victor eventually said,

"You give incentive if the product is of low quality. Have you ever given incentives?" Bosco was equally irate. He was expecting Papa to talk. He did not like to talk to the college boy. However, he said,

"I was asking Papa, not you."

Papa walked as if he did not hear the exchange between Bosco and Victor. Bosco asked loudly,

"Are you retiring after your bumper harvest?"

Papa ignored answering. The snub made Bosco very angry. He walked away to go home. He was probably planning something bad, Papa thought.

11

Bosco could not bear to see the success of Papa and his grandson. The high crop yield and the quick sale of the entire crop was more than all he could bear. He was curious driven by envy; he had to know how the fortunes of Papa were on the rise. In the annual gathering of farmers after harvesting, Bosco made a gesture of friendship by inviting Papa and Mama.
Victor and Titti had to attend the invitation from Bosco to represent Papa and Mama.
Mama attempted to educate Victor and Titti about Bosco's background. She wanted her grandson to prepare before the meeting with the crafty neighbour.
"The parents of Bosco arrived, accompanied their son to buy one failing farm. In a hurry, Bosco bought two adjacent farms making his farm large enough for grazing sheep and pigs. He behaved friendly in the beginning as he approached many farmers intending to buy their farms. The news of a man with cash to splash and willing to pay over the odds for those willing to sell circulated among farmers. Papa was aware of such practices when rich folks would pretend to buy farms to sell later as prime land for development. He imagined that they would build many polluting factories to avoid the stricter urban controls. He called a meeting of his neighbours and all agreed to put a stop to Bosco's land-grabbing expansion ambitions.
Papa became the number one enemy as Bosco learned of the group's opposition to his ability to purchase any other land in the farming valley. The tension between Papa and Bosco started as Bosco saw Papa as his stumbling block towards expansion."
Papa said,
"At times, he used his animals as weapons to destroy crops. When he noted winemaking was profitable he planted vines and

offered to buy Mama's grapes proposing a high price to stop her from brewing and Mama agreed."
"Why did you agree to such an arrangement?" Victor asked Mama.
"He did many underhand practices like stealing the workers by offering higher pay. That caused disruption in the brewing process. We chose to sell the grapes, not to please him, but to protect the farm from reduced profits."
Mama whispered to Titti and Victor saying,
"Please tell me how his wines taste if he offered. Be aware that you will surely get problems with Bosco, no matter how nice he speaks to you. He is not a good man and you know he is bitter that we are brewing again."
"Do not speak evil of your host but enjoy the hospitality," said Papa.
"I can tell him when I think he is wrong, can't I?" asked Victor.
"I leave that to your choice. Ensure you avoid trouble," said Papa.
"There will be no trouble from us, Mama," said Titti.

Aware of the situation, Victor and Titti left to go to Bosco's bash. There were no smiles as they approached. No one who saw Victor and Titti said a word such as "welcome." There was, however, an airy silence when Victor and Titti entered the large barn. They did not recognise many of the faces. There were no friends of Papa. That silence, the staring and the shaking of heads continued until Bosco spoke. He started with a statement that changed to a question saying,
"It is customary that we, the farmers, meet socially after harvest to exchange ideas. What is your opinion?"
"It is a good idea. I am happy to know that," replied Victor.
The keen ear of Titti heard Bosco whispering,
"There is a lot you will have to learn boy."

Victor and Titti mingled and pretended to drink the cheap bland wine. Titti whispered to Victor,

"Who will buy this sour drink? It has more vinegar than discarded fruit juice."

Victor said,

"Just be nice for now, a single word of criticism will start a clash."

Soon Victor stopped smiling to answer a question. It was from a friend of Bosco who appeared intoxicated.

"We hear you are a college boy. Are you going to continue with your grandad's farm or are you aspiring to take a high-flyer city job?"

Victor was not happy when the man called him a boy in front of his wife. He was offended but remained silent expecting a repeat but it did not happen. There was tension. The silence enraged one fat farmer, who asked,

"Your grandfather is getting old, is that the reason he sells his crops cheap?" Victor was grinding his teeth. He saw trouble. He had to diffuse the situation but he did not know how to do it. His grandfather had asked him to be nice to the host. However, the hostility was too much to tolerate. He confronted the unfriendly questioning by saying,

"I am the farmer now. I sell if the price offered is profitable for me. I produce quality and I get prices fitting my quality. That is what I learned in college."

"How was it possible that you could sell all your products but we could not?" said one fat farmer, sitting on a settee drinking beer from a large mug.

"You pay me and I will teach you," said Victor.

Bosco was vexed on hearing the conceited response.

He asked,

"Do you think you are our teacher? Can you lecture us on farming after just one season of work?"

"I don't intend to lecture anyone but give answers if asked," said Victor calmly.

Bosco smirked and said,
"You mean to lecture on how to sell cheap."
"I do not sell cheap, you are telling me." Victor was emphatic.
"Prove it," Bosco said.
Victor faced Bosco, grinned and said,
"It is better to ask the buyers."
"Who are your buyers?" Bosco asked abruptly.
"Ask them for the prices because it is the same buyers we all use. You will not believe hearing from me."
Bosco moved scornfully to grab a glass of wine which he drank in large gulps before saying,
"Your buyers are from the city. They are not our usual buyers from the town. Give us their names."
Victor noted Bosco's desperate attempts to know. He smiled and said,
"Ask all these people, they saw the logos on the trucks, please contact them. You are assuming they bought cheaply from me. Would you want to sell cheap to those who bought cheaply from me?"
There were murmurs as no one could recall noticing any logos. Some were not sure whether the old man sold his food crops cheaply or not.
Bosco was afraid of losing face in front of his friends. He pointed a finger at Victor and said,
"You college boys are liars. I will discuss this matter with the old man."
"You are welcome," Victor replied.
Titti moved closer to Victor, faced Bosco and said in a bold voice,
"Have you any evidence that our prices are low?" Bosco looked at Titti and then at Victor and said,
"Yes, you sell cheap. I saw a van collecting wine. You must sell cheaper because as beginners you could not beat my quality. How else could you sell all your wine?"

"We are still in trials. Those buyers thought it was good and wanted to see how the market reacts," said Titti.
"You should have brought some bottles for sampling. We could grade your quality," said Bosco.
"We did not know that this party was for wine grading. We will take part next time," said Titti. Bosco looked at Victor and Titti with obvious anger. Titti signalled Victor to leave.
"Your cheap wine would not qualify for entry in this bash," said Bosco after a brief silence.
"Our wine is sold at a higher price at the farm. You must have noticed that our wine is more expensive in the shops than yours. Your wine does not sell although it is cheaper in the shops. What does it tell you?" said Titti giving Bosco a challenge. Bosco was silent thinking of a venomous comment he could say in response when Victor interrupted that moment of silence and said,
"You know how businesses are conducted. It is not on prices only but do not forget quality." Bosco was too angry to talk as Victor and Titti put on the table half-full glasses of their drinks and approached the door to go home.
Victor turned and said to Bosco,
"You better stop making wine, if you cannot compete, start planting potatoes to feed your pigs and sheep."
Bosco moved to try to hit Victor. He threw a punch that missed landing on Victor. Bosco fell. He may have punched too forcefully. He uttered angrily some profanities as some friends helped him to stand up. He must have consumed too many tumblers of alcoholic beverages. Victor and Titti heard the insults but did not give any response. They proceeded to walk to go home. As they walked, Victor and Titti thought about what Papa and Mama could say:

The situation was a lesson
A glimpse of a future to fear
A spur to start protective measures

Papa listened to the unfriendly talk that occurred at the gathering and said,
"Bosco is ignorant, materially rich but poor mentally and that makes him resort to solving problems through violence."
"They were hostile to us from the start," said Titti.
Papa nodded as he said,
"You are young, clever and successful. Bosco is envious, bitter and unintelligent. His inability to buy this land has made him irrational. Now you pose a greater threat to his expansion plans."
"Do you think he could use violence try to undermine our progress?" asked Victor.
"Not when I am alive, as he knows I have more friends than him," said Papa.
"What could he do anyway?" asked Titti.
"He could set fire to our silos. Be friendly but watch him. We will respond in a manner he could not understand," said Mama.
"Do not be afraid to expand you're your farming business," said Papa.
"Do not declare your plans. Let them ask you and that will prove they are watching. Make peace with him. Remember you insulted him when you told him to plant potatoes," said Mama.
"I do not know how or whether I can appease him," said Victor.
"Tell him about your plans to plant potatoes and ask for his advice or opinion," said Papa.
"Bosco is proud; your humble approach may calm him. He will not forget the insult but may lower the hostility," said Mama.
Victor thought as he rubbed his hair. He made a bold decision to visit Bosco.

12

"I am sorry for acting in a rude manner last night," said Victor.
"I was expecting an apology or else I would demand it," said Bosco.
"Thank you," said Victor.
"Now what do you want?"
"I do not need anything. I was undisciplined. Mama said it is not good to say a bad thing to your host."
Bosco grinned and said,
"Now I know you are Mama's boy. Are you ready to sell?" Victor looked at Bosco sternly and said,
"That is Papa's business not for me to decide. I work hard to make the farm productive. You should support me not fight with me because I believe that hard work pays."
Bosco smiled and said,
"I saw people at your farm. What did they want?"
"They came to talk about new crops and suggested potatoes."
"Is that all?"
"I plan to extend my farm by planting potatoes," said Victor.
Bosco laughed menacingly and commented,
"Expanding to plant potatoes? What a daft idea. It is a waste of time, energy and money. Is it your idea or that of the old man?"
"It is my idea. I will try for one season only," said Victor.
"I cannot wish you luck for such foolish plans but I see you expanding your vines. Did they tell you, in addition, to plant grape-vines?" Bosco asked.
"Titti is experimenting, she studied winemaking in college."
Bosco clenched his fists, turned away and said,
"You are out to disrupt my wine business. I will not let you do that."
Victor left.

A Dot on the Timeline

The abrupt ending of the visit for peace was an undeclared conflict in the making like when two nations are on the verge of war. Nations on the verge of war are aware that conflicts cause lots of misery and suffering but instead of diffusing the tensions, they talk to show power. In this case, it appeared that Victor preached peace while Bosco preached intent of domination. It was unlikely that an amicable existence could occur.

Victor arrived home shaken but not put off by his plans of farm expansion and said to Mama,

"I learned that Bosco is still thinking of buying our farm. He was most upset when I told him that Titti has winemaking knowledge."

The news of the disagreement between Bosco and Victor travelled as fast as fire in a dry wild forest on a windy day. Goodwood visited Papa early in the morning. He was sure that Bosco was capable of causing harm. He did not want Papa or his grandson to fall victim to Bosco's wickedness.

In the days that followed, bitter exchanges between Bosco and Victor were escalating day by day. Papa noted Bosco walking aimlessly around his farm and at times talking alone. Papa told Victor,

"It is demeaning and pathetic to see a man talking to himself."

Victor commented,

"Bosco looks like a man in need of support rather than a man to fear. Shall I suggest buying his farm from him?" Papa laughed aloud went toward Mama and said,

"Can you imagine Victor buying Bosco's farm?"

Mama laughed aloud and said,

"If he were to sell to you, I will jump and catch a fly in flight."

Victor laughed and said,

"I was only joking."

"It is not a joke, you may be able to make that happen, Try it," said Papa muttering,

"That idea is worth a celebration."

Victor and Papa, on their daily early morning inspection of the farm, found the newly planted vines uprooted and scattered on the farm. Stray animals walked through the farm. There were clear clues that the animals that did the damage were the sheep from Bosco's farm. Victor, in his temper, shed a tear. Without asking Papa, Victor went to see Bosco, called him out to see the damage and asked,

"Did your sheep cross into my farm?"

"Oh yes, there was nothing growing and they could not do any crop damage."

"Did you ask for permission?" asked Victor.

"I don't need any permission as your grandfather had allowed me in the past," said Bosco with a smirk.

"There was crop damage of young vines. Your sheep trampled and uprooted the young shoots as they started to give buds. From now on, you ask me, before your sheep cross into my boundary," said Victor.

"I do not have to ask you because soon you will sell. You better take the money and vamoose," said Bosco defiantly as he turned away from Victor and walked off. Victor called Bosco's name very loudly and in a tone of authority saying,

"Bosco, stop and listen. You are the one to flee. I have plans to buy your farm. You better start packing."

No one had threatened Bosco before. He was shaken, but was furious as he said,

"I will burn your farm to the ground before I make a run for it."

"You do that I will burn your sheep in a bonfire. You will smell the burning of the wool even if you run to the other side of the globe. To start I will kill any sheep that crosses this boundary into my farm. I will fill the belly with potato stuffing and feed it to your pigs."

Bosco walked slowly to his home in silence, like a dog with its tail between its legs. He was thinking about how to answer or retaliate. He did not want to argue or show his fear to Victor, the young, strong and well-educated man. Papa was an earshot away when Victor shouted at Bosco. He heard the threat and could not believe his grandson had such a temper. He kept that discovery to himself but later he was tempted to tell Mama.

In Days Past

13

Victor left. He was angry.
After a moment of calm, he thought:

> *Anger is not a good emotion*
> *Even though, it may strengthen the resolve to win*
> *But may encourage thoughts of violence while*
> *Blocking the line of thoughts for peace*

Papa sat quietly grinding his teeth as he listened to Victor's lament. Victor, close to tears was telling Papa his quarrels with Bosco including an exchange of threats he made to Bosco, forcing him to walk away angry and in silence.

Papa had enough of Bosco. He decided to tell his grandson what Goodwood had told him about Bosco's early childhood behaviour.

"A childhood schoolmate of Bosco with a farm on the north side has also experienced unsocial activities from him. The neighbour, whose name is Goodwood, recognised him at your wedding and was sure Bosco did not recognise him. Bosco bullied weaker and younger children at school. He failed the exams to go to senior school. After many trials in many schools and in desperation, his father asked him what he wanted to do as an adult. Can you guess what the reply was? He wanted to get a lot of money by working in the city. His father allowed him to go to the city to try his luck for a job. No one knows what happened in the town. It was rumoured that Bosco returned home penniless, malnourished and dressed in rags. The heartbroken parents bought him the first farm because he said he could not fail to rear sheep. No one could recall any sighting the parents visiting their son at the farm. Goodwood noted that

Bosco suddenly became very arrogant with money to splash. Goodwood found out the reason for the sudden behaviour change. Bosco's parents had died and from the inherited money, he became proud and arrogance. He intimidated some to sell to enlarge his farm."
Papa concluded by saying to Victor,
"Now you know the rest of his activities."
Victor rubbed his hair, rather hard and vigorously. He thought about how a bully like Bosco could have parents who supported him through the thick and thin periods of life. He remembered his parents and wondered what would happen if they were alive. Papa, Mama, and Titti listened to Victor's fury. It was not normal for Victor to utter profanities, but with hate, he directed his insults to Bosco. In tears, he asked Papa.
"Tell me anything about my parents' character you can remember."
Papa stood up and in silence went to his bedroom. Mama looked at Victor and said,
"He remembers the accident. It is still painful and cannot talk about your parents without a tear falling. He was our only son and you are our only grandson. Please just give him time, he will be back to talk to you."
Papa returned and said,
"Your father died at the age you are now. I was just getting to know him as a man." Papa stretched his hand to give Victor a nicely folded very old brown envelope. Victor thought it was something from his parents and his body started to shake. Titti noted Victor's condition, took the envelope, opened it and showed Victor two separate pieces of newspaper cuttings. One was large and the other was very small. She said,
"Shall I read?" Victor was unable to speak. He nodded and listened. Before the reading Mama said,
"Papa had never dared to read details about the accident. Please Titti read slowly for him to grasp."

Titti asked Victor,
"Are you ready?" Victor nodded again and Titti started reading the larger cutting that gave the details of the accident. Victor remembered it. Mama had told him about that incident on the day he learned about the parents' fatal accident. Titti paused, looked at Victor and then opened the small piece to read; it was an obituary.

"Charlene our adopted daughter,
Rest in peace"

Vince and Josephine signed the obituary.
Papa stopped the rocking of his chair and cried out,
"What?"
Mama stood up and said,
"Child, read it again. Titti read the note again very slowly to ensure there was no mistake.

We did not know that Charlene was an adopted child. We met her parents but they disappeared without a trace and we lost contact before they said anything about the adoption. We are not sure whether they were alive or dead. We read that obituary but we surely, in that time of stress, we missed reading it properly."
Victor asked Titti to read more but that was the end of the article. Victor looked at Papa and Mama and then said,
"Surely there must be a grandmother and grandad I did not know. Can I try to find my mother's biological parents?"
Victor's mind was heavy with many thoughts.

Would the knowing bring tranquillity or fear?
Could it influence the future?
Why is my life with so many unknowns?

14

Papa and Mama were restless. They were probably guilty and ashamed. Missing crucial writing in the newspaper those many years ago was disturbing. They admitted their error and promised to help to find Charlene's biological mother and father, the biological maternal grandparents of Victor. There were no clues such as names or residential addresses. Nearly fifty years had passed; the search could be very difficult. Mama said to Victor, "We are sorry. In the early days, we were struggling and the death of your parents added to our misery. In distress, we could not repeat reading that newspaper cutting again. We thought of looking after you as our most cherished act."
Victor, close to tears said to Mama,
"Do not say sorry to me. You did the best and I will always respect that."
Victor asked,
"How was life in those early days?"
Mama called Titti to sit close to her and then said,
"Do you want to know how we acquired the farm?
"Yes, please Mama go on and tell us," said Titti.
Mama got a nod from Papa to tell Victor and Titti their struggles in life as an attempt to show Victor how difficult their past lives had been. Papa and Mama thought that telling the story of tussles would prepare Victor and Titti for any future difficulties and ease their emotions brought about by the information from that newspaper cutting.
Mama licked her dry lips and said,
"We lived in the city doing odd jobs until it was too hard to survive. We could afford to eat once a day, in addition, we were unable to pay our rent. In desperation, we moved to the farm areas looking for farm work that was likely to provide

accommodation. We did every type of work that a farm could offer, accepting work from any farm. We worked hard and one boss appreciated our work and decided to keep us for a longer period. After some seasons, we learned he was in debt and planned to sell his large farm. Papa asked his boss whether it was possible to buy a small piece of land. We planned to cultivate it just to get food to survive."

"That was like my parents. Our farm was very small but able to provide food for us," Titti interrupted. Mama resumed her tale of the past and said,

"Papa had no money. I wondered where he could get money. Up to now, I cannot understand where your grandad got the courage to suggest such a request. The boss suggested something pleasing. He showed Papa a rugged piece of land with boulders like an ancient riverbed, impossible to cultivate using machinery. He thought no one planning modern farming would buy that land. When Papa heard the price, he said,

"Boss I will work for you free for six months. That could pay for it." I cringed. I thought the man was going doolally. The boss went away saying he was thinking. I asked Papa how we could afford to get food without a wage. He said,

"The boss will have to supply food to enable us to get the energy to work."

The boss took a day to think and then he said,

"I am going to the city to find a buyer; in the meantime, you take care of the farm." Papa agreed. He had the whole farm to manage without knowing the answer to his request. Papa and I were temporary farmers behaving as bosses.

We worked as hard as before and we had a good harvest, which we sold and saved money for the boss. The farm provided all the food we needed. As the planting season approached, we were worried that someone could come and evict us. We had no place to go. Could we claim the small piece of land the boss showed

A Dot on the Timeline

us? These thoughts troubled us. Our worries increased as time passed.

One day, a smart young man, dressed in a suit, came to the farm driving a very clean expensive car. He said,

"My father is very ill. He did not find a buyer but you can use the farm, pay him every month and after ten years, the farm will be yours." It was an offer like no other. Papa and I saw the amount written on a small piece of paper. It was a large sum to pay. I scratched him before he signed and said,

"Where the hell will you get the money?" Papa looked at me smiled and said,

"We will sell the wood in the land he had given us. We plant fast crops and sell them in the market and with time, we will find more money. It is a good chance of becoming owners, as we will not get a better opportunity. He was good to us. He is ill we cannot let him down."

Papa signed the contract, shaking, sweating and worried that failure to pay may result in losing everything. With hard work, good favourable weather and at times bumper harvests, Papa was able to make payments in full and on time. In the land that the boss had given to Papa for the services rendered, Papa had planted trees to show that the land belonged to someone to avoid people trespassing. In the years that followed, the demand for timber was high. House building was at its peak. Building firms needed timber in large quantities. The price for timber rose and anyone with trees for timber was likely to get rich. The trees were valuable and generated funds. The payments became easier and we were able to build this house, your father was then a young adult thinking of getting married. He worked hard on this farm. I could never sell this farm. When he died, I vowed to preserve his memory on this farm for his son for the future, as you were only three years old. When Bosco came flashy and boasting that he could buy our lands I was active in stopping him. He will never pardon me for my actions and I will not apologise."

"Now young man, let us find your past and build a future," said Papa after listening and enjoying the account of the past from Mama. Victor close to tears stood and embraced Mama and Papa saying,
"Thank you for preserving that past. You are the best grandparents."
"We will ensure we make it better for the future," said Titti.
Papa patted Victor's back and said,
"I think the situation calls for a bit of relaxation. A bottle of Titti's wine might just do that."
"I think Titti sold all the wine bottles," said Mama.

15

Titti stood up as a person protesting in a debate and said, "You taught me one thing, Mama."

"I taught you many things Titti, not one thing," said Mama playfully.

"You told me to have enough produce for family use for a season, and there were no exceptions and my wine was on that list," said Titti.

"I am proud you listened," said Mama smiling.

After a sip or two of the wine, Victor asked,

"You made Titti a very good brewer. She is now as good as you are. How did you get into winemaking Mama?"

"It was a long time ago. It started purely by accident. It was not something I had thought to do in any of our farming plans. A young man, who looked malnourished, with torn dirty clothes came to me as I was working on the farm. He reminded me of our early days with Papa. He was carrying a basket containing plants. I thought they were flowers. I had some money but could not afford to spend it on flowers. He stopped and greeted me with great politeness and said,

"Please buy these. I cannot carry them anymore because I have not eaten for two days." I took pity on the hungry boy and gave him the snack I was to eat for lunch. The speed at which he ate prompted me to give him my juice drink. He sat for a while and then said,

"Can I help you to cultivate?" That request moved me emotionally. I nearly asked him to stay with us but we were too poor to afford to take care of him. I offered to buy his shrubs, although I did not know if they were of any value. He smiled and said that I had already paid by saving his life and he promised to bring more shrubs. I was curious why the boy was selling useless

shrubs. I asked if the shrubs had any food use. He smiled again and said that they produce sweet juicy berries that were edible and when squeezed produce sweet fruit juice. I planted the shrubs and they flourished and we realised they were grape vines. I regularly made grape juice. We made a lot more as the number of fruits increased. We stored some in glass bottles in the outer house and forgot. One day I heard an explosive sound coming from the shed. I called Papa to accompany me to investigate. We opened the door. The shed was wet, sprayed with red paint but smelling like a brewery. Papa saw the red colour and said,
"Your juice bottles have exploded." The smell of beer made me ask Papa,
"Did you secretly store your beer bottled here? You said you cannot afford to buy alcohol." Papa swore that he had no beer in the shed. We had to know the cause of the explosion. We went to see a neighbour, a sheep farmer who made cheese. The lesson he gave us converted me from a juice maker to a brewer of wines and I became the best. I remembered and wished to see the boy who gave me the shrubs but I never saw him again."

Titti was happy as Mama concluded her story. She vowed to preserve the knowledge for Alena. Victor was probably thinking about his maternal grandmother and grandfather and did not listen carefully to Mama and Titti's talk because he did not give any comment.
Titti realised that Victor was in deep thought and a cheering story could uplift his mood. She looked at Papa and Mama. She was happy to see how relaxed they were after the stories of their past. She remembered one crucial question. She asked,

"How did you two lovely people meet?" Papa was about to doze off because of the wine but Mama was alert and said,

"Do not ask Papa. He may pretend not to remember or he truly does not recall. Do not worry my child; I remember the day as if it was yesterday."
Victor showed interest and said,
"We would be proud to hear that." Mama moved closer to Victor and Titti and said,
"I was in a shop buying some food. It was rice. I had tried many food shops and they had no good quality rice. When the man weighed the kilogram of rice and put it in the bag, it looked very small. I said to the grocer,
"It looks smaller, have you weighed correctly?"
The man said,
"This is good quality rice it is heavier so it looks smaller compared to other cheaper quality products." There was a young man behind me waiting for his turn. He stepped forward and said loudly to the grocer,
"Let me check this weight with another scale at the other shop."
"Go ahead," said the grocer.
The young man picked the bag up. He was about to take the bag away when the grocer said,
"You pay me first."
"No, either you correct this error or we call the authorities. This young lady should get exactly what she is paying for."
The grocer said,
"Let me check again."
The young man said,
"You should apologise to the young lady. Let her take that bag of rice. Weigh a new bag correctly for her to pay. If you agree, that will be the end of the argument."
More people heard the argument, entered the shop and asked,
"What is happening?" The grocer acted quickly to the demand to avoid uproar as we left the shop quickly. While we were outside, I had two bags of rice and I offered to take the smaller bag when he said,

"You are a better cook; I will come for dinner instead." Since then, he has been my protector and he never forgot that first dinner."
Papa said,
"Since then, you have made improvements in your cooking."
Victor and Titti proceeded to go to bed smiling.

16

In the weeks that followed, Papa and Mama expected Victor to calm down. They thought the story of their past could calm him. Unknown to them Victor was in agony, thinking of possibilities and the likely hurdles he could encounter in finding his maternal grandparents. He had to act. He remembered a colleague who could help. Victor contacted him.
It was Bevan.
Bevan had amazing search talents. He was capable of finding anything lost, hidden or stolen. He displayed those skills prominently in their outdoor games while in college together. In finding hidden treasures, Bevan proved the best. He could notice hidden clues better than anyone could. Victor smiled as he picked up the phone to give him a call. Victor spoke slowly in sadness, hoping Bevan would take pity and accept the task. Bevan had only the newspaper cutting that had details of the accident and the obituary lines to start his research to find Victor's history.
Bevan, to calm his distressed friend said,
"Yes, I will search. I cannot promise but I will try to get a solution, my friend. You stay put and leave the matter to me."
Victor breathed a sigh of relief as he replaced the phone handle with its cradle.
Bevan had very scanty information as a starting point. The date of the accident and the age of Charlene, Victor's mother, became the little information that helped Bevan to start searching. Victor was joyous but not excited because the chances of success were slim due to the long time that had lapsed and the little information supplied.
A week is a long time when waiting for news. A month could be an agonizingly stretched period. Victor had to bear that long period of waiting. Bosco's intimidating acts did not stop but the

importance of the news Victor was waiting for took priority. The absence of walks on the farm and the silence that the family maintained for a period that lasted over a month became a matter of concern to Bosco who questioned whether Papa was ill. Bosco did not get any response to his inquisitive questions or his menacing incidents. He became more agitated and frustrated. He was in fear that Victor may carry on the threat of forcing him out of his farm, especially in the absence of the old man. His fear made him alert and in a constant state of observation. He sent friends to find out what was making the family go silent. Bosco's behaviour of walking around the edge of the farm made Papa, Victor, Mama and Titti think that Bosco was planning to do something wicked. The family and all farm workers were at maximum alert. It was a period when the family experienced tension from waiting and from an unhinged neighbour.

Victor thought that Bevan did not find anything and was afraid to return with news of failure. He was starting to despair at the thought of dismissing the chance of knowing his past. He started to follow his farm routine to let the past rest. He could not erase from his mind the wish to find the maternal grandparents. Vince and Josephine were probably dead and they could not help. These thoughts could be responsible for Victor's subdued mood and affected his ability to make or take jokes.

Victor woke up early to go to inspect the farm. He opened his door to find Papa waiting. Great minds think alike, both must have thought about that famous saying.

There was a potato patch at the end of the farm. Victor noted some anomalies and looked at the plot angrily. Bosco's sheep had walked on exposing some newly planted seeds. He picked one germinating tuber with green buds. He threw it to the ground in disgust and turned away from his grandad to ensure the old man did not hear, some profanities directed at Bosco. He started sweating and in a temper, walking towards Bosco's house. Papa noticed Victor's body language and said,

A Dot on the Timeline

"Do not go until you cool your temper down."
Before he could answer to express his anger to his grandad, he heard Titti calling excitingly and aloud,
"You have a visitor."
A call like the ring tone of a phone or sight of a visitor attracts speedy attention.
"A visitor, did you say?" Victor shouted thinking who the visitor could be. The family had rarely any guests or visitors and that surprised Victor. Victor walked fast keen to see and hear the intention of the visitor. He arrived at the house almost breathless. When he noted it was Bevan, his heart rate increased further giving him increased sweat. Without saying a word, he pulled a chair and sat. Bevan noted that Victor was restless. He stood, shook Victor's hand, smiled and then said,
"Do not worry, there is some progress. After a detailed search, Sofia is the name of your biological maternal grandmother."
At that moment, Victor was as attentive as a hunting fox. He smiled for a short time. He had a name that marked his past, a spot on the timeline of his family. He then heard Bevan saying,
"She gave birth to twins a boy and a girl. The twins named Charles and Charlene were adapted to different families as she was unable to care for them or maybe, it was due to the stigma put on a girl bearing children out of wedlock in those days of the past.
I will need more time to find the whereabouts of the twin brother, Charles and your grandmother, Sofia."
"I will be glad to meet them," said Victor, almost in a whisper, after clearing his throat and licking his dry lips.
Bevan left saying,
"I will continue the search but as there are possible name changes in adoptions, my search will be harder but there is hope."

17

Victor thanked Bevan and trusted his findings to be authentic without asking him how he came to that conclusion. The hunt for Charles and Sofia was on Bevan's mind as he left. Victor was very pleased. He had no pictures of his mother but the name Sofia linked him to a past. The name Sofia was enough to make him whistle his nursery rhymes cheerily as he returned to the farm. He forgot all about what he was going to do to Bosco. The name Sofia made Victor happy. The enormity of the information was taking its time to sink in. He was jovial as he reported that talk with Bevan to Papa and Mama who were proud to hear of Sofia and looked forward to a day they could probably meet.

After knowing of Sofia, another set of happy moments happened. The birth of another child added to the family glory by rejuvenating Papa and Mama much more. For another four weeks, the family maintained a private existence. Victor and Titti decided to name the child Sofia in respect of Victor's maternal grandmother. Happiness in the family had no bounds.

Bosco did not get any response for his unsocial acts on the farm. Ignoring Bosco was a punishment he could not bear because he behaved strangely. Bosco was walking, talking and at times shouting profanities. The walking and talking subsided when someone told him of the birth of Sofia. That did not make him happy but quite the reverse occurred. Any birth in Papa's household threatened his chances of gaining control of the farm. He probably wished for something nasty to happen to Sofia.

Weeks passed which became months or more. No one was counting. The period did not appear to be too long as the family had other exciting issues of birth and rejoicing that masked the passing of time. Sofia was only a baby but had the power to make the old rejuvenated and the young excited while making Bosco in danger of getting into a mental crisis worrying of losing the chance of acquiring Papa's farm.

18

Victor was eager to see Bevan. Each day that passed, his wish to see Bevan became more intense. Like a child expecting the return of the mother from shopping, Victor constantly peeped through the window at the path that led to the house and one day he saw Bevan approaching. Fear gripped Victor who did not know how to react. He was in a dilemma at the sight of Bevan walking slowly. He was worried that the information Bevan had, could be good, bad or different from the previous information.

The sight of a brown envelope handed to Victor meant there were serious issues, increasing the fear and the question of why Bevan did not just start a conversation.

Bevan did not show any emotion as he said,

"You will find details in the envelope. That is not all; there is more to do. Give me a call when you have checked the names."

Bevan stretched his hand for a handshake after which he left. He did not stay, as he did not have answers to the whereabouts of Sofia. The speed of delivery of the news made Victor very nervous. All the family members gathered around Victor as they looked at Bevan disappearing away from the farmhouse. Victor held the brown envelope, thought for a while, and remembered that so far in his life, brown envelopes delivered important news. It was for either very good or very bad news but nothing else. With shaking hands, Victor tried unsuccessfully to open the brown envelope. His shaky hands were then sweaty. He asked Titti to help to open the letter blaming the firm glue. Titti took the letter from the envelope and handed it to Victor. She feared reading it. She feared delivering gloomy news, she thought. Victor read it very quickly and then in a shaky loud tone of voice, he read it again.

A Dot on the Timeline

> *"My friend, the search found three people*
> *With the name Charles, that matches your mum's birthday.*
> *In addition, I matched those names to addresses shown*
> *Regarding Sofia, there is no report.*
> *However, all the searches indicate that Sofia is not alive*
> *But I need to confirm."*

Victor looked up as if he did not understand the letter. He read it again, the third time. The shaking increased, the more he looked at the addresses that matched the name Charles, the more his tremors manifested. Titti noticed the agitation and asked, "What is the matter?"

Victor paused, wiped his sweat and then shouted, "What is this? It cannot be correct. I think Bevan is wrong." Titti was concerned and in a worried tone of voice she asked again, "What is wrong?

Papa and Mama listened to every word said or read. There was a moment of silence when Victor eventually said,

"One of the addresses matches that of Bosco." The choking voice of Victor made the grandparents' jaws drop, stunned and unable to ask any questions. They knew it was very bad news.

It was a frightful thought to all in the family that Charles could be Bosco, the family nemesis.

Titti said,

"It is impossible, a nice man like you cannot have a bully, a cheat with malice and all that is bad about a man to be your uncle. It cannot be."

Victor in a very weak voice said to Titti,

"My dear wife, I dread that possibility too. What shall I do if it is true?"

"Can we refuse to acknowledge him, Papa?" Titti asked in desperation.
Mama said,
"Let us stop thinking too far. Papa, you should go to Bosco with Victor. There is very little chance that he could be our Charles. The sooner you do that the better. It will be the best thing to do to eliminate this woeful thought."
Papa shook his head and said,
"How shall I get answers from that rude, mentally deranged man?"
Mama looked at Papa sternly and said,
"You always know what to do. Ask him a series of questions to see whether his answers match the information you have in this letter."
"What shall we do if it is him?" Papa asked.
"You will know what to do as always. The situation will guide you. However, we must know for sure. I cannot spend a night not knowing," said Mama assertively.
Papa agreed to escort Victor saying,
"We should calm down. We will go in the evening after planning our questions."
Papa tried to make sense of the situation by taking a stroll to enable a quiet moment of thinking. He went through the trees hoping the fresh air in the trees and the aroma of flowers in the meadows could help in bringing thoughts of wisdom.
Victor went for a walk through the grape vines, thinking that the fragrance of the ripening fruits could give him solace.
Mama and Titti decided to calm the stress by trying a new recipe for a starter for the evening meal.
Alena and Sofia appear to sense the tension by playing without crying.
Victor returned without answers to the acute problem. He entered the house. He walked round and round like a cat chasing its tail. In thinking of suitable questions to ask Bosco, Victor met

a brick wall. He did not have a single question he could ask, without giving clues about their intention. It was a total failure. It was getting late; he had failed and had to rely on Papa to find the truth from a liar.

Victor saw Papa walking slowly returning to the house. He looked calm as he said,

"Victor, this is the brown envelope you did not want to receive. It is not for you alone but also for all of us. We face the issue together."

19

Papa's heart was beating faster as he watched the setting sun. Oddly, he did not enjoy the beauty of the setting sun as before nor did he note when the sun disappeared in the horizon although he was sitting and looking at it. His thoughts, however, were on Charles.
Victor had imagined Charles to be kind and loving like Papa and Mama. The thought of talking to Bosco as the likely uncle called Charles, made Victor's heart sink. Papa and Victor were in a conundrum.
They did not know what to do if he were the uncle or not.
Titti thought that, in most cases, a journey to find a lost relative would be a happy occasion. However, in this case, they were terrified if they found their Charles. Titti looked at Papa and Victor, as they got ready for their mission. She was in teary eyes when she said,
"Good luck. I wish you do not find Charles in that house."
It was a sad situation indeed. Papa and Victor realising it was dark, hurriedly walked to see Bosco. They took their bright hurricane lamp and a flashlight to ensure Bosco saw their approach.
Papa said to Victor,
"We must be visible from a distance as Bosco is on a war path. Before we step on his land we ensure we see him first."
"I cannot imagine him as my uncle," said Victor.
"I cannot imagine him as a family member, a friend or a person I could ever respect."
They were within the view of Bosco's house. Victor shone his flashlight to see Bosco holding a machete as he stood by his veranda watching the approach of Papa and Victor.
Victor said,

"He has a weapon Papa, I am empty-handed."
"Calm down we are going in peace," said Papa.
"Why is he smiling Papa?" asked Victor.
"That is not a smile. It is a grin like that of a dog before a fight or a bite. He may assume we are going to burn his house as you told him but he is probably ready for you. Look at his sword. It is not shiny. It may be very blunt," said Papa as a joke. Victor was not amused; he knew that a blunt machete could cause injury or fatality.
Fear made him shout,
"We come in peace, no violence please."
Bosco relaxed and put his machete on the wooden table, smiled and said,
"You come at this late hour, it must be important. If you came to buy this farm, forget it and go back. Approach only if you came to talk about a sale of your land or an apology for your insults."
"We have come for an open talk that could include many possibilities," said Victor.
"What kind of talk is that?" asked Bosco.
Victor quickly moved near Papa's ear and whispered,
"Help me here Papa I do not know what else to say."
Papa said to Bosco,
"How can we talk about business standing?"
"I will bring extra chairs, we can sit on the veranda," said Bosco in an elated mood.
Papa and Victor sat opposite Bosco separated by his dusty wooden table. They kept their eyes firmly focused on Bosco. They kept silent.

20

After a brief moment, Bosco was agitated as he asked,
"Did you come to waste my time by sitting here in silence or did you come to talk?"
"We came to talk but our throats are very dry, we need some water or some wine," said Papa.
"I think you came to tease me. You are getting neither wine nor water until you talk," said Bosco showing signs of anger.
"Is that how you welcome your guests?" asked Papa.
"You are not my guests, you are callers that are about to outstay their welcome," said Bosco standing behind his chair.
"We are guests not visitors and we are not strangers. You must treat us well if you want to hear from us. We go it will be your loss," said Papa looking very serious. Victor looked at Papa and thought that he wanted to abandon the talks. He touched Papa's hand and squeezed it. There was an airy silence. Bosco resumed a sitting position and asked,
"What brought you here? If it is business let us talk."
Papa smiled, he had the upper hand as he said,
"How can we sell if we do not know your full name and a bit of your background?" asked Papa.
Bosco, standing again as a person about to give a talk asked,
"Is that all?"
Papa nodded, he did not want to state a lie.
Bosco said,
"My name is Charles Bosco Justin. My adoptive parents named me Bosco. Charles was the birth name that I rarely use."
"Why don't you use the name Charles?" asked Papa.
"That woman who gave me away named me Charles. I don't want that name," responded Bosco with his mouth twisted then looked down as if he was going to cry.

Papa and Victor looked at each other knowing for sure that was their man. Victor's heart was beating fast with his legs wobbly. He held to the table to avoid revealing his shaky hands. The sweaty hands marked the dusty wooden table but the lamp was not sufficiently bright to reveal the imprint. Papa pulled his chair closer to the table to hide his shaky hands that he placed on his thighs away from sight and then said to Bosco,
"Please sit down."
Bosco looked at Papa and Victor from head to chest as the rest of the body was under the table as they were sitting. He grimaced and then said,
"Why should I sit down? I can stand if I want. This is my house. I did not tell you to sit, did I?"
"Don't be rude to my grandad, he can sit if he wants," said Victor in a tone of voice that was intimidating to Bosco.
Papa quickly spoke to stop Bosco from answering Victor,
"We need a glass of wine. It is important in talking about important issues."
"What kind of sick joke is this? When did you start to be this friendly? To come to my house at night to ask for wine is sick even to friends."
Bosco said that shaking in temper.
Papa was sure he had the upper hand. He said,
"We have come to talk about serious matters. We want the wine, the best of what you have. We will not drink cheap wine."
It appeared that Papa had convinced Bosco that the business involved a sale. Bosco said,
"It is not nice to drink before a business; you may sign the wrong document and I don't drink at these late hours."
Papa said,
"Have you got the wine or not? You do not have to sign documents today. An agreement will allow drinking today and signing tomorrow."

"What type of agreement?" Bosco asked showing signs of lessening in tension.

"You will drink more than one tumbler when you hear what we have to say," said Papa as he moved the chair intending to make himself more comfortably seated giving Bosco a hint that he was expecting the wine.

Bosco hesitantly went to the back room. Victor looked at Papa and nodded to show approval of the progress. Papa and Victor continued to maintain silence to avoid giving a hint of their revelations.

Bosco was unsure of how to react. He was still in doubt about the outcome of the meeting. He was not smiling when he arrived with three tumblers and a bottle of red wine from Titti's brewery. He made sure that Papa and Victor saw the label. After pouring the wine, Bosco continued to look at Papa suspiciously in silence and in anticipation of what Papa was offering. He sat down keen to listen. Papa took a sip. His mouth was dry in anticipation of what he was about to reveal. Victor did likewise to ease his shaky, sweaty hands and dry mouth. Bosco drank to ease his nerves. He took a sip and said that the wine was very nice. They did not expect that approval, as an enemy does not give a positive comment to a rival. Papa and Victor were certain that Bosco was thinking of the sale. Papa gave a false hint of a sale plan when he said,

"It is a very nice wine; you will enjoy it as yours in the future." Bosco could not hide his eagerness and with a smile, he responded by saying,

"I will certainly enjoy it."

"I am sure you will," Papa said looking directly at Bosco, who paid full attention in anticipation to what Papa was about to say, "There is one question I want to ask before we conclude." Bosco quickly replied saying,

"Please ask anything."

"Did your adoptive parents tell you the name of your biological parents?" Bosco put his glass down on the wooden table with a thud and said,
"Is this an inquisition? Are you insulting me in my own house? Please go. I am tired of this type of sales tactics."
"Calm down, just answer what you know and all will be revealed," said Papa in a calm tone of voice indicating he was serious.
It appeared that Bosco saw the serious nature of the question as he promptly replied,
"I was told she was called Sofia,"
"Is she alive?" Victor asked.
"No, that is why I am alone. All my parents are dead. I have no one in this world related to me. I am in reality alone and independent of anyone."
He stood up and pointed a finger to Papa and Victor shaking it from side to side and then said,
"Neither the old nor the young would frighten me." He sat down for a moment and then asked,
"Why such questions, I do not understand. I have never in my fifty years of life talked about my real mother," said Bosco nearly in tears.
Victor correlated all that Bevan reported and found a complete match. He started to wonder how he would get on with such an uncle.
Papa looked at Bosco, took pity on him as he portrayed humility and said,
"We are nearly there; just answer this question which may make things clear to you.
Bosco nodded and said,
"What more harm can you do? Go on and ask but if it is another insult, I will ask you to go."
"Did anyone tell you about a woman called Charlene?" asked Papa.

"I do not know such a person," said Bosco affirmatively.
"Have you heard the name mentioned by anyone?" Papa asked.
No, I have not heard such a name," said Bosco standing to let Papa and Victor out.
"Take a sip of wine," said Papa.
Victor filled Bosco's tumbler with wine and gave him. Bosco took the tumbler, and held it in his hand but refused to drink saying,
"Why are you asking all these hurting questions? You came here to tease and insult me. You have succeeded in making me sad with your false belief of offering a sale. Now go."
Papa, in an authoritative tone of voice, said,
"You are not alone. Take a glass of wine and drink it. Charlene was your twin sister who died in a car accident leaving a son."
Bosco looked up, and then he understood what Papa had said. His glass slipped and fell to the floor smashing.
"Don't worry, your nephew will clean it, said Papa. Bosco stared at Victor but did not find words to say. After a sip of the wine from Victor's glass, Bosco asked,
"How did you find out all that information? Why did they hide that from me, why no one talked to me and why…" and he started crying. Papa paused and allowed Bosco to cry. Papa almost cried when he saw Victor in tears too.
Papa emptied the bottle of wine, as there was a lot of crying taking place. Victor could not address Bosco as uncle the feelings of hostility were too raw. Bosco could not mention the word nephew. Both words were not easy to utter. Time was the only hope that could allow healing to take place.
To make the awkward situation bearable, Papa addressed Bosco as a son and said,
"It was one incident that led to another and it became a long story that we will tell you tomorrow," said Papa as he stood to walk home. Bosco continued crying until Papa and Victor left.

A Dot on the Timeline

Papa and Victor were walking slowly exchanging their feelings that Bosco was their man when they heard, "Wait for me please." Victor turned to see Bosco approaching with his powerful solar lantern. He stopped to breathe and said, "Please let me come to talk with you for a longer period." In silence, the three men walked to meet Mama and Titti. They could not hold arms as bosom friends would do. They walked in silence. To Bosco, however, that was the best company he ever had. Victor was in shock thinking of the immediate issues of reconciliation for which he had to find courage and answers.

I cannot utter the word uncle
I do not want to hear nephew
Here is the uncle
Ugly as things may be
What do I do? The past is set
I did not choose to be a nephew

The sight that Mama and Titti saw, Bosco walking between Papa and Victor, was sufficient to confirm that they found Charles. They could not close their mouths in disbelief. They remained silent waiting to hear the verdict from Papa and Victor. Papa and Victor did not need to tell of the outcome. It was clear from the humble behaviour of Bosco. He sat at the table to try to eat the dinner offered to him. The tension needed a cup of tea or a bottle of wine. The traumatic discovery of an uncle and likewise a nephew needed calm surroundings. To understand and accept the situation, Papa had to think of something to bring about relaxation. He knew there were many questions to which he could not find answers. The silence at the table needed something mollifying to allow the information to sink in. It was only Papa or Mama who could find something appealing to bring tranquillity.

Papa sat on his rocking chair. The silence was deafening. Mama moved closer to Papa and whispered saying,
"Tell them one of the many childhood stories to bring some laughter."
Papa replied,
"Fat chance, a story will not do. We need a miracle." Mama looked sternly at Papa, a sign for him to do something and fast. She said,
"A story with miracles could help. Can you recall one from your stories?"

21

You can either hate it or love it but you cannot ignore the ripples that a brown envelope can create. It was not easy to describe the situation that the brown envelope caused. It was an understatement to call the situation a surprise. It was more like a shockwave that travelled to cause effect far and beyond the source. The situation needed calming down. The drastic change in relationships would have to take time to sink in. Papa was in shock too but as a leader, he had to prevent the damage that shocks can cause. To calm emotions Papa decided to tell a story as suggested by Mama. It was a story, told not for its wit, fun or entertainment but with the role of diverting thoughts of hatred to those of love, sadness to happiness or agitation to serenity. The narration was probably a method of coping with the emotion caused by the discovery of Bosco who was Charles. Although the verdict of finding Charles was not a hundred percent certain, at that time a peaceful co-existence was desirable. Bosco was as silent as a hunting owl. He was in shock to find the past he knew was very incomplete. Bosco's emotions were mixed and raw. He feared whether he could cope with what he had just then found out. He could not talk. His facial expression showed a timid demeanour although he had some satisfaction of belonging to a family. He was ready to listen to Papa's story keenly from start to finish. He was unsure his newly acquired family would love him. He was ashamed of his recent acts. He was thankful to Papa who was forgiving. That was his family, how could he face them? He wished for time to help in the healing.

Victor had not spoken for a while. In his mind were many thoughts that kept on repeating as a stuck record. The questions in his mind kept him in a dazed mood. To try to make sense of all that happened he kept on reciting:

> *I know my past now*
> *Terrible as it may be but it is mine*
> *I need strength to cope with what I know*
> *To prepare for what is to come*

The murderous path that Bosco and Victor were in the process of creating had to change. The saying that time heals and the new family bond could probably form to bring reconciliation was hard to believe. That thought in that early stage of creating the new family could not sit easily with Victor and probably Papa, Mama and Titti and that would be the same in Bosco's mind.

At first, Mama was in a state of denial. After a good while of thinking, she knew she had only one path to follow and that was to accept him. Titti thought she would take a while to change her abhorrence of Bosco to a loving uncle. She thought it was not going to be easy. She tried to, express her feelings to Victor but he did not speak to comment, as his feelings were too raw and painful to allow discussion.

Papa took a sip of wine as he relaxed swaying slowly on his rocking chair and looking at Bosco and Victor to try to detect any common family features like nose, eyes or lips that linked them. Sofia was breastfeeding while Alena was sitting on Mama's lap talking. Unexpectedly Alena said to Mama.

"My mummy promised to buy a dog."

"Did you promise her a real dog for Christmas?" Mama asked Titti.

"Yes, Mama, a small one," Titti answered.

Papa was tired and about to sleep. He became alarmed when he heard Titti's answer. He asked Alena,

"Do you want a real live dog?"

A Dot on the Timeline

"Calm down Papa. Not everyone is afraid of dogs like you," said Mama.

"Is Papa afraid of a little dog?" asked Alena.

"Sit and hold me tight, we will ask him to tell us why he fears dogs."

Papa coughed to clear his throat and said,

"I was planning to tell you a story about my school days but a dog story will do."

"When I was a little boy," Papa paused as Alena laughed very loudly.

"Why are you laughing?" asked Mama.

"He is not a little boy," Alena replied.

"He was a little boy once many years ago and I was a little girl as you are now," answered Mama.

Alena smiled as she looked at her parents for reassurance as she listened to her great-grandfather's story. Victor and Bosco, too traumatised to say anything, sat to hear the dog's story.

22

"Kimani was a neighbour who lived alone. He was selfish but rich and strongly disliked the poor. One day, from his walks visiting his rich friends, he had a dog. Kimani started walking while talking to the dog in profanities that referred to how he disliked the poor. The bizarre change in Kimani's lifestyle caused concern about his state of mind. Many were curious about the change in character and wondered whether dogs were good or bad for the mind. Kimani looked down on the poor as disgusting beings similar to vermin, fit to flush into sewers. The poor beings, in turn, wished him nothing pleasing. It was not long before many rich people had dogs on leashes and talking to dogs was a means of showing that a person had riches. There were more dogs in the village than people. Many especially the poor hated dogs and their masters. Feeding was costly. They associated dogs with eating rotten meat and dead bodies. Many lips whispered that dog ownership was a curse and wished the owners a bad omen to rid them of their wealth.

Food shortages became critical when rains failed for several years and that situation resulted in people fearing hunger that caused a general unrest. There was an increase in theft and unsocial behaviours. The belief in the curse increased in acceptance. The presence of dogs, as the cause of the disaster, added to the theory of the curse.

The high number of poorly fed dogs started to behave like their wild species in hunting, constantly snarling in temper and stealing food. The dogs started biting people and snatching their food.

"I was a victim of a dog bite," said Papa as he pulled up his trouser leg to reveal the big scar on his leg. Alena was very scared. She held Mama more tightly and closed her eyes.

"Please Papa cover the scar, the little girl is scared," said Mama.

The hungry dogs travelled far to forage by day and at night they abandoned their guard posts. The dog owners locked the dogs in kernels during the day and released them at night for guard duties, however, the dogs did not behave as expected. They absconded to go to forage at nights. That widespread act of caging the dogs, apart from adding to their suffering and stress, had undesirable and unexpected impacts.

On one moonlit night, the dogs met in the valley to scavenge. TD, the top dog, called for attention with his loud bark. It was a full moon, they barked more than usual. It was obvious the dogs were unhappy about the behaviour of their masters. TD asked, "What do we do to get a better deal from the humans?"
There was a chorus of voices saying,
"We attack them, bite them or kill their young."
Other dogs said,
"We make them fear us by behaving like our wild cousins on hunting. We can become disobedient and violent or we can pretend not to understand their commands."
"They have tools to kill us, chain us or they could refuse to offer the scrapes we get," said TD.
"We run away to find new pastures," said one big dog.
"Do we want to go to the dogs' home to be in cages again?" one little dog asked.
"We are locked in cages in the day, what is the difference?" an old dog small in stature asked.
"At least we get better treatment at the dogs' home," said one brown spotted dog.
TD said,
"Stop arguing. Freedom comes with responsibilities. Tasks like finding food in the wild are hard. With humans, we get food and a home for good but we get bad treatment. We fight to get better treatment,"
"How?" asked the small dog.

No dog responded. No dog had an answer and there was a pause in the exchanges until TD said,
"Let us talk to them in their language,"
All the dogs barked once and then kept silent. That meant it was a risky act.
Musa, a very wise old dog said,
"It is too dangerous,"
"Why?" asked the little dog.
"The humans would be like us. They will be plagued by the *curse of the ghost of dogs* and it is not known how we would feel with them among us," said Musa.
TD said,
"The humans would feel the suffering we endure."
"The suffering is unbearable. It is better to do it and all will face the consequences. If things go badly for us we will undo it," said the brown spotted dog.
One big but malnourished dog barked and said,
"How can you remove that curse of the ghost of dogs from the humans?"
There was silence and TD wanted to see progress in their mission. He turned to Musa and said,
"Please help. You are old and wise."
"There is a secret password," said Musa.
"Please tell us," said TD.
"You never tell the password until the appointed day," said Musa.
"When is that day?" asked TD.
"On the night of the full moon you should meet and the password will be revealed. I will give you a clue; it consists of two words," said Musa.
Papa paused to take a sip of wine. It was a chance to look at the faces of Victor and Bosco to see if they were calming down. He saw two gloomy faces gazing at the sky. He knew he needed more ideas to help bring about calm. Alena called,

"Papa, do dogs talk?"
"Dogs communicate in a language humans do not understand," said Mama.
Papa was happy that Mama helped. He coughed, put his wine glass on the stool next to his rocking chair and continued with the story.
In the morning, Kimani opened the door to find TD waiting and snarling as usual.
"Move," he said to the dog. Kimani was surprised to hear, "Good morning Kimani." He thought a neighbour had visited. He moved forward to see who spoke but kicked the dog hard. He looked at the dog; he saw the dog speaking in a deep tone of voice saying,
"Kimani, look down before you move."
A ghost entered Kimani's mind. In his trance, he walked fast like a dog following a scent. He saw other village people running away. He followed them. They met. They did not know what they were doing and they could not speak. They sat together in silence surrounded by their dogs that were silent too. The affected people had lost the power of speech and the ability to think. Their minds were like those of the dogs or maybe worse. That is what the curse of the ghost of dogs does to people making their lives unnatural. Papa after a pause to breathe, he then said,
"The curse of the ghost of dogs was frightening to the poor who had no dogs and witnessed the great change of behaviour of the dog owners."
The dogs left to go foraging in the valley of carcasses, followed by their masters. It was a land depression like a quarry. People threw into it all the dead animals. The severe water shortage caused many deaths to humans and beasts alike.
There is a saying:
"A loss to one is gain to another."

The valley was a food source for the dogs. Some villagers with sound minds followed curiously intending to know what was happening.

They saw Kimani fighting with some dogs over the rotting leg of a dead donkey. He had snatched it from the dogs and was eating. There was no doubt he was deranged. The fear of the dogs and the crazy people increased. Fear forced many to lock their doors and erect higher and stronger fences to prevent entry by dogs or afflicted humans. The cursed people smelt like rotten flesh and snarled like dogs causing great fear in the village.

A search for a cure commenced. Finding a cure could make a poor person rich.

The village folk were waiting for the cure to end while the dogs waited for the time of the full moon. It was in the interest of the dogs to reverse the curse. To the poor and the sane human, finding a cure for the curse of the ghost of dogs could mean riches. The witch doctors of the village were busy searching for a cure and that could make them famous and rich.

Food availability had changed from bad to worse. The dogs needed humans to provide food and the sane humans were busy finding a cure to be rich.

The night of the full moon was approaching. All the dogs assembled at the valley of carcasses. Barking at the full moon, they expected Musa to arrive. No one knew the whereabouts of Musa, the wise old dog with the password. A frantic search revealed that Musa had died before the full moon. The top dog, TD, did not have the password but a clue that Musa gave him.

"It is two words that humans say to one another," said TD.

Desperately there was a list of suggestions. It took a long time to find a reasonable pair of words. A small dog, that had lived in the dog's home for a long time went close to TD and whispered,

"My name is SD. At the dogs' home, I always heard humans telling one another *'Merry Christmas'* and they replied the same.

At the same time, many dogs were barking continuously at the huge moon that shone brightly as it is now."
TD frolicked and said,
"It was the most likely password. We speak those two words early in the morning."
A chorus of barking signalled a unanimous agreement. All the dogs bowed in front of TD as they barked and trotted to go home.
In the morning, TD woke up and barked the password to his master. Kimani woke up; he suddenly spoke and said,
"Get out of my bed." Without the demons, his full mind and power of speech restored, Kimani realised he had slept in the dog's house. He panicked, crawled out and ran to his neighbours shouting repeatedly, "*Merry Christmas*, it is Christmas today." He could not remember what had happened before. He was not aware of his stench and filth. He stood singing "*Merry Christmas*" happily many times, joined by the others, released from the bond of the curse of the ghost of dogs. After a wash and shave, Kimani planned a celebration for all, the rich and the poor, not forgetting a delicacy to his dogs. That was a great day of happiness in the village. It was Christmas. It became a day to remember as the end of the curse that caused fear and misery in the village and a day to show virtues and to suppress vices."
Papa reclined on his settee and finished his wine from the glass. Bosco sat quietly with hands folded, hardly moving.
Victor was reflective on the story:

The end of a tragedy was delight
Is Christmas a mark on a timeline?
A day for start of happiness

After a short time of silence, enough for Victor, Bosco and Titti to take in the meaning of the dog story, Papa was already asleep.

Victor, Mama and Titti had considered Bosco an archenemy as recently as before that evening meal. He had been an enemy since they met him and that was many years ago. Sitting and sipping wine from a glass tumbler on the veranda of Papa's house looked strange to all and possibly to him. The silence continued as Bosco could not laugh, make or take a joke as his mind was laden with too many thoughts. He was glad for the welcome by a family he nearly destroyed by self-indulgence.

Family Concerns

23

Victor and probably Papa, Mama including Titti, were reacting to the traumatic events of discovering that an archenemy was an uncle. Until that day, Victor had no past family outside Papa and Mama. When he discovered a family member, he was in a dilemma. He needed help out of that predicament.
The humility shown by Bosco in realising he belonged to a family started to influence Victor's thoughts. He had hated Bosco but could not turn him away. Something Papa had said a long time ago came back to his memory:

> *The family bond is an asset to cherish*
> *Lose it; life takes a route to perish*
> *Find it; hold it, life will flourish*
> *As precious as dinner with added relish*

It was nearing bedtime but no one was ready to go to sleep. Papa woke up from his short nap finding everyone seated by the table and wondered why they were so silent.
He said to Mama,
"The mind takes time to process information, especially if it is shocking in nature."
Mama said,
"The whole family is in shock. We have to trade carefully until the time the silence ends."

To the family, mainly Bosco and Victor, that single day marked a dot on the timeline their cycle of life. However, the shock needed a longer period to understand and accept as reality.

A Dot on the Timeline

It started like any other day, a brown envelope received brought with it thoughts of fear, doubt, worry and disbelief marking an indelible spot on a time of life of the family.
After a long time of thinking, Mama had to break up the silence by saying something mollifying by saying,
"With time the fear and the worries will wear off to give rise to hilarious stories that could be passed through the generations as folk tales. The thoughts of this day will become indelible in your memories." Victor coughed to try to respond to Mama's comments. After several attempts he said,
"It is easier said than done. I cannot see how to get out of this shock. Please Mama we need help."
Mama saw Bosco nodding and said,
"Papa's story talks of situations that no one had any escape plan. Time however provided the means that led to happiness for all. We hope that will be same in our case."
Bosco and Victor nodded in agreement and said nothing.

Bosco never thought that his feud with Victor would end until he got what he wanted. Time, however, healed the rift through an incident that Bosco or Victor could never have predicted. The sudden shift from loneliness to having someone calling him uncle was a shock to Bosco and it was to Victor an unusual act for someone to call him nephew.
According to Papa's story, Kimani never thought he would thrust aside all his wealth and eat dog food or go to sleep in a dog kennel. At Christmas when he recovered, he realised his pride and the belittling of others was brief and nothing but time was able to wipe away the conceit.
It was getting late in the night. Papa was supposed to go to bed. He did not feel the urge to do that. He glanced at Mama, Victor, Titti, Alena, Sofia and Bosco sitting. They were nibbling snacks silently without a hint of hostility and he said,
"Cherish this moment and hope the future is always pleasing."

Bosco coughed several times before he could speak to say,
"From loneliness to join a family of six is wealth greater than any I have ever owned. I thought I was rich but I was a poor man thinking more of wealth than people."
Victor managed to smile hearing such sweet words from a man he could soon call uncle and said,
"The cutting from a 30-year-old newspaper changed many things. It was like emerging from a dry hot desert in misery to a cool, refreshing and pleasing power shower."
Mama indicated that she wanted to say something by standing and lifting her wine glass to take a sip. Instead of speaking, she took time looking at Alena, Sofia, Victor and Bosco. She was trying to find any matching family traits. She was still too traumatic to see but she believed she would find something significant in time. She held Alena and Sofia to face Bosco and said,
"This is your great uncle." Bosco stood up and tenderly held the hands of Alena and Sofia then looking at Victor, with teary eyes he said,
"Please Victor let us go to my house for a few minutes." Victor obliged without hesitation.

The bedroom was huge. It was full of items in such a disorganised manner that it looked like a junkyard. Victor did not comment on what he saw. He could not afford another stress that day. He was learning how to respect his uncle. Bosco did a quick search and he found something that he put in his pocket. They returned quickly without any conversation.
From his pocket, he took two small gold bracelets and put them on the hand of each child and with tears in his eyes, he said,
"My adoptive parents said I was wearing these when I arrived."
Mama's mouth was open for too long that Titti noticed and said,
"What is the matter, Mama?"

Mama stood up and held Titti's hand. They walked fast and went to the attic together. Mama did not say anything but started searching. She opened a small box and said,
"Yes, I found it." They rushed to show Papa who nodded with approval. Mama faced Bosco and Victor and then she said, "These bracelets belonged to Charlene, your mother." Mama handed them to Victor. Victor held them and cried copiously as he kissed the bracelets. She allowed Victor a moment to calm down before handing them to Bosco. Bosco received the bracelets, sobbing like a baby in need of a cuddle from mummy. Mama looked at Bosco and allowed him to calm down before saying,
"Your sister had these in her jewellery box. I preserved them when she died."
Bosco looked at the bracelets and cried again. They were similar to those he had put on the children's hands. He stood and put the bracelets on each child's hand again.
The four small baby gold bracelets were identical in colour size and markings.
It was irrefutable as proof that Charlene and Bosco (Charles) were twins, a brother and a sister.
No one takes it lightly when grown-up men cry and sob like children. That was what Bosco and Victor did. Victor for the first time stood and hugged Bosco as they cried like children. Papa and Mama looked on with teary eyes. They had done their crying thirty years ago. Bosco was still sobbing when he said,
"Please Papa and Mama, can I stay here tonight? I do not want to go to that big house alone. I do not want to be away and alone from my family again."
Mama noted the humble side of Bosco and allowed him to stay in Papa's spare bedroom. Victor led Bosco into the bedroom and said,

"Good night." Bosco did not know what to say as no one had said such a good thing to him before but he coughed to get his voice and said,
"Have a good rest my nephew."
You could guess that Bosco could not fall asleep. Bosco thought of the events of the day as he lay on the bed in Papa's spare room. What dominated his mind was how a day can dramatically change a man's life and wondered how futile it is to predict the future.
Victor likewise thought of the future as he lay on the bed talking about his feelings to Titti. The children slept happily unaware of the events that occurred but were proud of a man Mama had called a great uncle who gave them gold bracelets.

24

The family that woke up the next day had increased from six to seven. You could make a wrong guess and think that the addition was a newly born baby but it was a fifty-year-old man. At the table, as they sat for breakfast, no one had previously tallied their titles. Victor was a grandson, a father, and a husband and then he became a nephew. Bosco was neither a father nor a husband but that morning he heard Papa and Mama giving him the respect of a son and Victor calling him an uncle. He smiled every time he heard those titles used. The tension and the hatred were melting away slowly like an ice cube in a glass of juice on a hot day.

Victor noted Bosco as he ate his breakfast silently looking like a person with guilt or shame. He ate very slowly and very little. In a timid voice, Bosco faced Papa and asked,

"Can Victor and I go to check the farms?" Papa and Mama were trying to understand the change in Bosco's behaviour. It was too much to take in within that short time. They noted the polite mannerism, the talk that showed respect and the sudden attachment to Victor. The polite attitude pleased them and they wished them a good first day together. They noted a hint of peace and love.

That first day of the new family was like any other with some subtle changes. There was no observed difference in the weather. The sun shone as usual and the sheep roamed as usual. To Bosco and Victor, it was a very different kind of day. They walked together in peace; they smiled not quarrelling and they exchanged pleasantries, not profanities. They talked about the farms as they walked about suggesting improvements and not destructions.

After a short walk, Bosco said,

"There is no doubt you are my nephew. First, I would like to address your grandparents as Papa and Mama. Secondly, I would like to take full responsibility for all damages I have done. Finally, I would like you and me to work together to make our farms great."

"It is a good idea. We will seek approval from Papa," said Victor. They walked hands shoulder to shoulder like friendly schoolchildren, a scene that did not escape the keen eyes of nosy neighbours.

They passed through the potato patch and Bosco witnessed the damage and said,

"Pardon me, nephew, I have no words to express my despicable acts, I will make amends."

"It is the family farm. You do what you have to do as you own it now. I thought that one day you would see the light, but the light has shone on you," said Victor.

"Indeed my nephew, indeed."

Papa and Mama, sitting on the veranda saw Bosco and Victor approaching. Their smiles as they chatted said it all.

Papa said to Mama,

"We are not a weak family anymore."

Mama did not answer as she was looking at something that attracted her attention and could not resist saying to Papa.

"Look at their teeth when they laugh and tell me what you see."

Victor and Bosco were close to Papa and Mama who scrutinised every inch of their face for clues of family traits. They hoped something unique would show up one day. Mama smiled at them as they approached and said,

"How did your farm inspection go?"

"We have a lot of work to do to correct the past and put our energies together for progress," said Bosco. The reply impressed Papa. He had not fully trusted Bosco. He thought that in time a few set tests could prove his genuine love for his newly found family. With that in mind Papa said,

A Dot on the Timeline

"I would like to hear the plans for the farms from the young minds after dinner this evening."
"I will have lots of requests Papa," replied Bosco.
Victor in smiles said,
"Bosco promised to plant more vines for Titti and the potatoes to correct the most recent damages. He also asks whether he can stay for longer in this house to contribute to looking after your welfare. He is afraid of going home and staying alone again."
Papa and Mama nodded with smiles as Bosco for the first time embraced Mama and shook Papa's hand saying,
"I will work hard to make you happy. Guide me to reach that goal."

Victor, in an elated mood, contacted Bevan. He gave him the good news that he found the uncle. He pleaded with him to do one more favour.
"Be patient and wait," said Bevan as he ended the phone call.
The tantalising wait ended after a week when Titti answered the phone from Bevan. Victor and Bosco sped to answer it and then they heard,
"There is not much to report because your grandmother died aged twenty-five after a short illness. There are no records of marriage. No one knew of her siblings as the writing on the gravestone was only her name."
"It means no one knew of our existence. We have to re-write the headstone," said Bosco sadly.
"We add the names of her children," said Victor.
"We have to visit the grave as soon as possible," said Bosco.

The recent family union that included Bosco was an aura of mystery to all the neighbours and friends. It was certain there would be questions. Papa's friends thought Bosco was buying the farm after noticing the frequency of visits. Goodwood met with Papa as matter of urgency to plead to Papa on behalf of many farmers, not to sell to Bosco. He was prepared to exceed Bosco's offer. Papa was curious. He knew the extent of hatred between the neighbours and Bosco but not the extremely wicked plans Bosco had if he acquired Papa's farm. He wanted to know more of what they thought of Bosco by asking,

"Why should I not sell to Bosco?" Goodwood stood still. He was surprised and worried that Papa could be intimidated to accept selling. He paused and then said in a begging tone,

"Please Papa do not sell. Bosco planned to cut off the source of irrigation water from the stream in your land. He would close the access road to some farms and let his sheep roam free in exchange for water. In the event he acquired your land, he will have the power to carry on his wicked ideas with impunity."

"That will never happen," said Papa in an authoritative stern voice.

"What will you do to prevent such acts?" asked Goodwood.

"He will not be able to do unsocial acts anymore," said Papa to the surprise of Goodwood.

"Please tell me that he is selling to you and he is going away," said Goodwood finding Papa's answers confusing and evasive.

"He is neither selling nor buying but I will assure you that he will behave." Goodwood was very happy and reassured that there would be peace. He looked at Papa curiously and asked,

"How have you managed to tame that aggressive bully?"

"I am not prepared to disclose all now but give us time to reflect on the changes that will soon be clear to all of us," said Papa reassuringly.

25

A mirage, you may think it is an object and then it disappears. Unlike a mirage, the sight of Victor and Bosco together frequently continued to remain a mystery to the neighbours but it was real. First, many farmers noted the end of hostile activities in their farms. That was a surprising change. The sight of two recent enemies walking, talking, laughing and at times showing friendships like patting each other's back, made many onlookers very curious by asking many questions. They could not imagine how Papa could tame the savage. The family had agreed to keep the discovery of an uncle a secret. A secret of more than one person is bound to leak but the strong bond of the family made the leak impossible despite many enquiries and temptations to reveal. It was a test of the strength of the family bond and Papa was happy that the secret remained a secret. The family was truly a unit, an entity that was gradually showing to stand the rigours of the past traumatic encounters from within and from outside influences.

The success of keeping the secret strengthened the family. Papa was beginning to trust Bosco and the family was growing stronger as a unit. Victor was building a rapport with his uncle and he was able to ask searching and personal questions.

One day after lunch, Victor asked,

"Uncle, I notice you love your great-nieces very much. You constantly shower presents on them. You must love children very much. Why did you choose not to have any?"

Bosco looked at Victor as he scratched the top of his scalp where the hair was receding at a rapid rate. His facial expression was that of a person reporting losing all his money and said,

"I have experienced rejection from many areas of life. The sudden death of my adoptive parents was as if I did not deserve my parents. The decision to send me for adoption by my mum always meant that I was not good enough for her. The failure to achieve in school, at work in the city and the inability to retain friends are fears that prevent me from bonding. In the event I had children, I would take it as a terrorist act by responding too violently for a mild disobedience, a misdemeanour or a naughty act.

Victor massaged his uncle's shoulders and with teary eyes he said, "You are not alone and you will never experience any more dismissals, rejections or thoughts of anger. You have the family for support."

"Thank you," said Bosco tearfully.

"Papa would say that our family is strong enough to suppress any emotions of anger," said Victor.

Bosco's revelation to his nephew about his innermost fear was a significant moment. Titti started to understand Bosco's past behaviour and explained to Victor saying,

"Now I know, Bosco concealed his fear and at times anger by boasting. There are remedial measures."

Victor dismissed Titti's comment as another ploy women use to show an understanding of emotions more than men do.

"The first step towards healing is by sharing a problem and you will note that all your problems will eventually disappear," Mama told Bosco when she heard of Bosco's innermost fears.

"I have my nephew, my great-nieces and you as parents. All my fears are gone," said Bosco smiling with relief that he was able to share his problems. Titti heard Bosco's comment that showed satisfaction and smiled but said nothing. When she saw Mama, she said,

"Did you hear what Bosco said?"

"He said that he is contented," replied Mama.

"Are you satisfied with that level of contentment?" asked Titti.

"I know what you mean. However, what is the solution?" asked Mama. Titti smiled again and said,
"Watch and see."
Mama smiled and said to Titti,
"Knowing you, your plans will fall into place and I will not wait for long."
Papa heard the exchange between Mama and Titti and said, "You plan too much. You must let time guide and heal."
Victor said,
"The women think they know how things are going to take place before they happen."
Titti said to Mama,
"I do not think the men know what we are talking about."
"They are holding the wrong end of the stick," said Mama. Titti looked at Mama and they broke into a raucous laughter.
Papa turned to face Bosco and said,
"You hear laughter and that is happiness in a family. A family bond is usually strong. It stretches in times of strife but it never breaks. You are now family and that bond link us all."
Bosco spoke emotionally to Papa saying,
"Now I feel protected, strong and never shall I bully anyone again. The knowing of my family has given me power to ensure the family bond will never stretch or break." Papa nodded and smiled. Bosco did likewise enabling Papa to see Bosco's teeth as he smiled. He noted the teeth trait similarities with Victor and confirmed Mama's observations. It was a comforting observation he shared with Mama. Papa made Bosco aware that he heard the promise he made by saying,
"Your promise is your bond. Do not stretch or sever it."
Bosco smiled as he nodded in agreement. Papa and Mama noticed the teeth again, smiled and giggled.
The family was exchanging views happily when the ringing of the telephone interrupted that moment. The sound of a ringing phone may bring about fear or pleasure but it always prompts a

swift answering. Victor and Bosco were jittery thinking of other surprises as Titti answered it.

26

"Who is it?"
"Sorry to disturb you, I am ringing looking for farm work," said a voice.
Titti paused for a while to think if she needed a worker. Before she had an answer, she heard,
"Hello, are you still there?"
"Yes, I was getting a pen to write some notes," said Titti.
Titti wrote fast and Sam spoke eloquently. She knew the age, the likes and dislikes, the educational level and finally the marital status.
"When can you come for an interview?
"Tomorrow," Sam replied without hesitation. Titti thought that Sam could help to tend her vines and help in brewing.
"When you arrive ask for Titti," Titti said as she put the phone down.
Victor and Bosco did not know the details of the conversation. Victor asked, "Who was on the phone?"
"Never mind, I may have a farm hand to help with my vines," said Titti.
"You do not need anyone now until harvesting," said Victor.
Titti surprised Victor and Bosco by smiling and saying,
"Grape vines need a daily tender loving care to give the wine that fruity and full-body you experience with pleasure when enjoying a glass."
Bosco said,
"It is a good idea; a farm hand will help us all."

Titti smiled when she heard Bosco and Victor planning. They spoke of tasks for the new employee. Cutting and carrying the poles for mending fences, are tasks that could be helpful. The leaking roofs of the barns and the broken barriers that needed constant repair could keep an employee very busy, they thought. They were looking forward to meeting the possible employee. Titti knew that Victor and Bosco were expecting a man with some manual skills and she said nothing.

At the veranda, Victor, Titti and Bosco were relaxing after an early breakfast, planning on the tasks for that day. Titti alone was expecting Sam. Victor saw a woman approaching their gate and alerted Titti. She was dressed in a suit like the city office workers. She wore high-heeled shoes that made her look awkward walking on the soft muddy path to the farmhouse. She opened the gate delicately touching it with her polished nails as if it were dirty. She wiped her hands with a brilliantly white handkerchief that she took from her shiny black handbag. She stood, maybe to rest her tired feet, looking at the farmhouse. She was probably debating whether she could work in that environment.
Titti stood up and said,
"Hello, what can I do for you?"
"I am Sam, you were expecting me and you must be Titti,"
Bosco and Victor could not close their open mouths until Titti nudged with her elbows.
"What do you do in the city?" Victor asked.
Before she could answer, Titti interrupted and said,
"The interview starts after a cup of tea." Victor and Bosco were wondering what to ask in that interview. Bosco was nervous and remained quiet. Sam sat and faced Victor and Bosco across the table as Titti was pouring the tea. The two men, however, were not thinking of an interview instead they were admiring the elegance of dress and sophistication of Sam's appearance. Her hair, her make-up and her earrings dazzled as the morning sun shone. However, they stopped gazing when they realised that Titti was looking at them as she sipped her tea and said,

"You were about to tell us what work you did in the town. When Sam spoke, she sounded like a person giving the interview. She showed scholarly sophistication in speech and language. She spoke eloquently in a calm voice, without hesitation as she answered saying,

"I was a turf accountant. I took the job temporarily as a gap period, waiting for a vacancy to work as a Veterinarian. Now I want to practice my training."

Titti was watching Bosco who had been very quiet. Bosco at that time was looking at Sam leaving the questioning to Victor, who said,

"This is Bosco, my uncle. He is likely to be your boss. He has a large flock of sheep and pigs. With a Veterinarian in his team, he may add chickens, goats, cattle and possibly horses and mules. Knowing my uncle, he may add donkeys to carry his crops to town."

All laughed and Sam said,

"I am a novice; I will improve but cannot claim to be an expert yet." Victor liked the genuine answer. He looked at Bosco for support and said,

"If you accept the terms of the job, you will work on his farm that we are going to sightsee with you."

Sam stood up and stretched her hand to Bosco for a handshake. The soft hand touch made Bosco tongue-tied as he nodded and led the way to his farm. Victor continued leading the interview. Before they left Titti said,

"Sam, those shoes are unsuitable in this environment, please try my boots."

Victor's interview was unique. He was looking at Sam's reaction. He passed near the farm boundary, intending for Sam to see the broken gates. He let Sam stand on a mound to grasp the spans of the farm and to see the grazing sheep. He passed near the barns close to the pig and sheep muck. He eventually went close to the house. Sam observed but spoke only to answer a question.

Victor conducted the tour of the sheep farm, by showing rather than explaining. When the tour ended, they stood in front of Bosco's house. Bosco found his voice to address Sam saying, "What is your opinion about the establishment?" Sam smiled. She turned and looked at the spans of the farm again. Bosco and Victor were biting their lips fearing rejection and then they heard, "I would be glad to work here but I would suggest substantial changes to increase profitability."
That reply was music to Bosco and Victor's ears. Victor imagines the potential Sam could bring. He decided to continue taking the lead in questioning and said,
"You will manage the farm, provide veterinary aid and make it profitable."
"That is fine," said Sam.
"The salary is not great; it will not be enough to pay for the fare to the town that is why I was suggesting accommodation at the farm," said Victor.
"That is fine," said Sam.
"We will go to see the accommodation," said Victor.
Bosco cringed with shame; the spare room in his house was very messy. Sam inspected the accommodation and came out smiling. Bosco smiled convinced that all was well.
"What is your verdict?" asked Bosco.
"Superb, I will fix the mess," said Sam.
Victor laughed.
"The manager's job is yours, should you choose to accept the pay," Bosco said looking directly at Sam. He saw the smile but hardly heard what she said.
"When do I start?" Sam repeated.
"At the beginning of next month," Bosco said.
"Excellent," replied Sam.

The interview concluded with smiles all round finalised by a vigorous shaking of hands. They returned to Papa's veranda. There were smiles of satisfaction.

Papa and Mama observed the conduct after the interview. Titti had prepared a snack. Her chat with Sam appeared very friendly and Sam did not show any urgency of leaving. Mama said to Papa,

"This interview has turned to look like a tour or a visit to friends." Papa and Mama got more surprises when Titti and Sam, after a moment of giggles, stood and walked to see the grape vines. They seem to strike up a friendship. The tour to visit the brewery was over when Titti went to Mama and said,

"Sam has some experience in brewing and she offered to help."

"Didn't she accept a job with Bosco?" Mama asked.

"She will start from tomorrow as a helper until she starts her real job at the beginning of the month," said Titti.

Mama looked at Titti's eyes to see whether there was a hidden motive and said,

"Her accommodation is not ready."

"She can use Papa's spare room and Bosco will choose what to do," said Titti.

Sam was happy and promised to return the next day and thanked Titti and Mama as she started to walk home.

27

"Sam, you are still wearing the farm boots, come and change."
"I will take the shoes to her, mammy," said Alena as she picked the high-heeled shoes she had admired.
Bosco approached running and said,
"I will pick you up. It is too wet for your little feet." Sam waited to receive her shoes held by Alena, seated on Bosco's shoulders. At earshot Bosco said,
"You can still wear the boots until you are away from the soft soils of the farm. You can bring the boots when you return."
"Thank you, boss," said Sam as she started walking away. Bosco was looking at her admiringly when Sam turned and waved.
Sitting on the veranda, Mama and Papa observed the farewell style and formed an opinion.
Papa said,
"Did you see how Bosco moved fast?"
"That was not an interview, but something is in Titti's mind. I will probe into her mind"
Victor observed Bosco's walk with Alena on his shoulders after the farewell to Sam. He thought there was a spring in Bosco's steps. He did not disclose that thought to anyone. Titti wondered how Bosco could trust an employee who could decide not to return despite a promise. She said to Bosco,
"You are getting too soft, uncle, you gave my boots away."
"I will buy a set of brand new boots for you and see whether Alena will have shoes like Sam," replied Bosco.
With a smile Titti said,
"She is too young for such glamour."
"Would you not like her to be trendy?" asked Victor.
Titti said,
"There is a time and stage for everything."

A Dot on the Timeline

It was the routine after the evening meal when the family discussed the events of the day. Papa and Mama sat quietly listening to the chat while glancing at Bosco who appeared to be in deep thoughts. Titti told the family about Sam's talents in brewing. Bosco was sipping a sample of wine that Titti had made and then suddenly he started laughing aloud. Titti, Mama, Victor and Papa turned to look at Bosco. They expected him to say something. They were proud but curious to know what made Bosco laugh so hilariously. He had not shown that type of behaviour before. It was not the norm. When Bosco noted that all eyes were on him he sipped some wine and said,

"In our farm we will have a good brewer and an animal doctor, what titles do we give to ourselves?"

No one had heard Bosco laughing, let alone making a joke. There was silence. Bosco must have been in a good mood. They wanted to know. Victor had to think fast to respond without spoiling that good mood. He quickly asked,

"Will the title of uncle and that of nephew be appropriate?"

Bosco smiled and said,

"That is the best title for you and me."

Victor asked again,

"Why are titles necessary?"

"Titles make anyone feel important. If you call Titti, MB (Master Brewer) she will always make the best wines in order to earn the respect of that title."

Victor laughed as he realised the humour of titles and said to Bosco that he would probably call his employee a doctor.

Papa and Mama could not resist laughing when Alena asked,

"What about me?"

Bosco stopped laughing and said,

"You are gold and Sofia is diamond. I will call you both precious."

Papa looked at his seven-member family that was acting as an entity and smiled with pride as he sipped the last drops of wine from the glass.

28

Mama's keen senses of observation surpassed any measuring scale. She could detect a subtle change of wind direction, the face of a liar trying to hide an undesirable deed or the guilt of a teenager with ideas of what they intended to do before they did. Victor remembered in his childhood that Mama could tell him to stop doing something he had not done but in hindsight, he realised that she observed his behaviour and predicted what he was going to do. She does not usually miss much when in a situation. There was a meaning to what she saw in the interview. Victor's wish was probably about to be granted, although Mama could not see how at that time. Victor wanted his uncle's farm to be great and that would make him happy. Titti's hunch was probably correct but she never gloated that she was sure it would turn out to be what happened next. Papa paid no attention because he was oblivion to what was happening around him or he did not care as long as there was peace and love in the family. Mama thought that something was pleasing about the employee; a thought she shared with Titti. Victor dismissed the thoughts of Mama and Titti as petty.

Mama was unsure of Bosco's awareness. She wanted to know whether Bosco was aware of the goings-on. She asked,
"When is your employee coming?"
"In a week's time, at the beginning of the month," said Bosco.
"Are you sure of that?" asked Mama.
"Not quite certain Mama," said Bosco.
"My son, remember you are the boss. You should know and control everything including the movement of your staff."
Bosco replied,
"Yes, I understand, Mama."
"What is your opinion on the new employee?" asked Mama.

"I think she is very clever and would be an asset to the farm," said Bosco.

"That is a modest assessment. Remember you have worked alone for too long. You must learn to share tasks and make your staff motivated," said Mama.

"I will learn as I go and will not disappoint you," said Bosco not fully grasping what Mama was saying.

A phone call ended the talk when Bosco rushed to answer it. He listened for a while and then said,

"OK."

Bosco was nervous. It was an invitation to all the farmers for a meeting to discuss issues of mutual interest. He called Victor and said,

"All those I have wronged will be there. What shall I do?"

"Do what you do best. Talk the least and look serious," said Victor to calm his uncle's fears.

"Is that what I do best?" asked Bosco.

"Yes, before I knew you were my uncle," said Victor smiling.

Bosco smiled too acknowledging it was a joke saying,

"Those were the dark days my nephew, please speak for me. I will be as quiet as an empty dance hall."

"I promise to speak to defend you. I will say that you had a vision. You saw a future of prosperity from a past you discovered. The vision prompted you to correct all the wrong things you did in the past."

Bosco nodded and said,

"I trust you nephew, because it is the truth said in a riddle. However, what will you say if they ask what kind of past did I discover? "

Victor smiled and said,

"I will say it was a dream you chose to follow."

Bosco smiling and shaking his head said,

"Thank goodness, I have a nephew. I will attend the meeting without fear."

The all-day meeting went well and Bosco escaped questions except for some glances, as Victor spoke. They walked home slowly when Victor suddenly stopped as they approached Papa's house.
"That must be Sam," said Victor.
"Oh, yes, I see her. She was not due to start work yet," said Bosco.
"You don't mind, do you uncle?"
"I do not mind at all, nephew. However, there is the issue of the accommodation. How can I make it suitable for her in such a short time?"
"She said she would tidy it up by herself. Let her face the challenge," said Victor smiling.
After greetings and a short talk about the meeting, Bosco asked Sam,
"You are early. Are you starting tomorrow?"
"I am on my holidays from the other job and Titti wanted to talk about vines and wines. I came to chat to Titti about new methods of extraction and fermentation," said Sam smiling.
Bosco's smile did not escape Mama's observation. Something pleased Mama about Sam. Bosco's frequent glances in the direction of Sam were probably why he forgot all about the cleaning of the accommodation as discussed with Victor.
The food, the chat and the opened bottle of wine caused all to surpass the time to go to bed.

29

A week does not appear a long time when you are very busy or very happy. That is what Sam realised as it was the next day that she was to start work on the sheep and pig farm. However, the lambing season did not give her a chance to have a second look and know the farm well. It started and it was frantic in its speed. It was the first test of Sam's ability as a manager and a Veterinarian. Bosco needed the family to help. He was dancing with joy to see Papa, Mama, Victor, Titti and the little girls Alena and Sofia ready and willing to help. The lambing that nearly doubled his sheep stock made Bosco very satisfied. He was proud of Sam's efficient work. Sam was proud of her work. She was confident when she suggested to Bosco to start milking the sheep to sell the milk to those who made cheese. Anything that would increase income received an agreement from Bosco. When Victor and Sam prepared and presented to Bosco the costs of the set-up, he was not as enthusiastic as before. He opted for wool production with cheaper setting-up costs.

Victor and Bosco spent time watching the flock of sheep and the little lambs in the fields as they made their daily morning inspections. It was a happy sight.
Sam was settling well in her room in Bosco's huge house. One evening as Sam was preparing her meals, Bosco entered the kitchen to congratulate her for the work well done. Instead of a handshake, it was an emotional hug. Could that be unprofessional? That was a thought in Bosco's mind, a boss and an employee to hug. It was only a harmless congratulatory hug, he thought.
"You can have a cup of chocolate to keep me company while I eat. I do not like to eat alone," said Sam. Those words rang a bell

in Bosco's mind. Before Bosco found out that he was an uncle, he lived alone, he cooked alone, and he ate alone. He discovered that it was not a good life when he discovered a nephew. He thought how sad it was and could not wish that situation on anyone. He accepted the drink and went on showering Sam with praises. The exchange of their life stories went on as she ate and as he sipped his chocolate drink. The talk made them more relaxed. They gained courage to suggest having breakfast together. When they had regular dinners together, they could no longer keep the bond a secret. It was like bonding in a team that enhanced work values. Bosco was becoming interested in the female company; the more they shared their life experiences.

30

The work of Sam, a very efficient employee, made Bosco a changed man.

Sam's story that hated a lone existence was significant to Bosco's lonely unhappy life. He had been alone for a long time. He was not alone as soon as he discovered a nephew. He liked that situation and hated the previous lone existence. He did not like to be alone and isolated like an island.

It did not take long for Victor, Titti, Papa and Mama to notice a change in Bosco. He was smiling more; he ate neither breakfast nor supper with them. After work, he sped to go to his house. Mama and Titti had a hunch, but Victor and Papa had to ask to know. Victor noticed a change in his uncle's behaviour when they were at work. The absent-mindedness was too intense. One day when they were repairing a broken gate, Bosco would lift a hammer and forget what he was going to do next until he asked his nephew. The constant smiles were obvious that Bosco had pleasing thoughts and that did not sit well with the level of forgetting he displayed. Victor had to ask,

"Uncle, your mind is not here, you are daydreaming of something pleasing. Is it your increasing wealth, or is there a secret you would rather keep private?"

"Nephew, it is not a secret, I was bursting to tell but I did not know how to start. I am becoming fond of Sam." Victor put his tools down, turned and patted Bosco's shoulder and said,

"That is amazing news. Tell me all and do not skip a single word."

Victor thought it was not going to be a short story because Bosco changed his mind and said,

"It is better to tell Papa and Mama."

A Dot on the Timeline

"Let's go to meet them as they will be as eager to hear the astonishing news as I am," said Victor walking fast.

Titti saw Victor and Bosco approaching the house smiling, laughing and slapping the back of each other in banter and thought something was exciting. Titti alerted Papa and Mama to what she saw happening between Bosco and Victor which she thought was very exciting. When Victor arrived he wasted no time but said,

"Bosco has something to say." Titti grabbed her daughters and went closer to hear the story. She thought that whatever they had to say was going to be pleasing because Victor and Bosco appeared happy and found it necessary to tell Papa and Mama at midday and not wait until evening. Bosco was still smiling as he started telling his personal story.

"I went into her kitchen to tell her how pleased I was with her work. She offered me a cup of hot cocoa. We started talking. That led to know each other much better. She had the same worries as me, the fear of loneliness and bonding like me. I offered to accompany her for breakfast and dinner so that she was not alone. One day after dinner, I went to my bedroom. My bedroom was changed. It looked clean and there were fewer jumbles. I could not sleep. I had to know what was happening. I had to go back to the kitchen where she was still busy tiding up and said,

"My bedroom looks different."

"Is it messy or is anything removed?" she asked. Without thinking, I said,

"It was too tidy. I never had it so good."

She laughed and then smiled. I was admiring that beautiful smile when I heard,

"I tidied it for you, my boss, I hope you like it."

I could not resist the compliment, I embraced her to say thank you and I kissed her cheek. I cannot fully remember the sequence of events that led to our lips touching. I got the courage and said,

"I love you, Sam." She repeated what I said and I held her closer. "You make me happy; I do not want to be alone again." She said the same words and I held her and we smooched."

Bosco paused to breathe. He saw Titti smiling and then he said, "That is why I have not been here to eat with you for some time. Last night we talked about taking the relationship further and she agreed."

"Including marriage?" Victor asked.

"Yes, my nephew, I did."

"What did she say?"

"Yes, she said yes." The smile revealed shiny clean teeth. Mama confirmed her family mark that linked Bosco to Victor. The teeth resemblance was as clear as a mirror image.

"Have you got a ring? I have your sister's rings. You can make it official," said Victor.

Titti sped to the farm to call Sam. Titti said to Sam, "We are having a special afternoon tea and Bosco wants you to be there." They arrived at Papa's veranda and sat quietly. Victor had given Bosco a quick short tuition on how to approach Sam with the ring. Bosco was a good student because Sam smiled and accepted the ring with the words: *"I do."* Papa asked Sam and Bosco,

"When will the great day take place?"

Bosco said,

"Papa, I have not thought about anything that far in the future. If the end of the month is suitable, we will be happy to fix the date."

Papa and Mama looked at each other and both nodded in agreement. Papa looked at Bosco and Sam and said,

"It is about a year since Sam came to the farming valley and she brought joy and happiness. The ring has added to the glee of Mama and I."

Titti did not waste time; she went and set a table. They sat enjoying an afternoon drink after a snack. The talk on wedding plans dominated the gathering at the table. The dresses, the decorations, the food and the booze were among the issues of discussion. The guest list was not long as it was mostly friends of Papa, Mama, Victor and Titti. Bosco and Sam had hardly any friends. Despite that level of few friends, Papa planned a big wedding.

31

Mama told them of the steps to take to make the celebrations unmatched. The talk went on for the rest of the day as measured by the number of wine bottles consumed.
Titti and Sam spoke of the wedding attire.
Bosco and Victor spoke about food and booze.
Papa and Mama spoke of the venue, the outlay and the guests.
Alena did not like it when no one gave her a task. She asked Sam, "What shall I do?" Sam looked at the little girl and then glanced at Titti to ask for an idea.
Titti said,
"You will be the bride's maids. You and Sofia will be smartly dressed to shower the bride with flowers at all times."
Mama was proud to have guessed correct that there was love in the air as soon as Sam arrived for the interview.
Victor's wish for his uncle's loneliness to disappear happened.
Titti was correct all the time. There was the possibility she set it up. Papa was simply happy that another marriage was taking place. Could that lead to a rapid expansion of his family? That thought was in the minds of all the adults but no one uttered a word. Sam did not think that her trip to the farm could transfer her fortunes from an employee to a bride and a wife.
Mama knew that any girl would be nervous about taking a bold step to join a new family. She called Sam for a chat to assess her feelings.

32

The ring on Sam's finger was significant. After the evening meal, Mama said to Sam,
"We never had a heart-to-heart talk between us."
Sam pulled the chair closer and away from the earshot of others and said,
"I was waiting for this chance to tell you about me and my family."
"I wanted to know if there is anything we could do to make you feel comfortable and loved," said Mama.
Sam smiled as she looked at Mama and then said,
"I am happy now but before I came here, I was a sad girl. As a coal miner's daughter, I witnessed the hard work my father had to do to support me in college."
Sam went on to describe how poor they were. Her father's wages supplemented by her mother's small income kept them afloat. An accident in the coalmine changed the family's fortunes. Disabled and unable to work, the family tumbled into poverty.
"My mother worked hard at times taking three jobs to sustain care for my father and enable my continuation with college. After graduation, I took my first job to reduce the struggles my mother had to endure to survive. The contributions helped to improve the health of my mother and my father's care was adequate. The improvement made me remain at home despite many offers of marriage."
"You are a loving child," said Mama.
"I came to the farm because I wanted to follow my profession. It would have more security and I could give better help to my parents. As an employee at the farm, my mother did not have to take more than one job to survive."
Sam paused to take a sip of the wine, and then asked,

"Is it a wise idea to move my parents from the dirty crime-ridden town to a dwelling in a farming community for them to enjoy clean air?" Mama replied and said,
"There are many unoccupied dwellings. They would be happy to reside near you."
"If that happens their quality of life would be much better and I will make it last forever," said Sam.
Mama was moved by Sam's story and said,
"You are here, you are family. Your parents are welcome to visit and if it is to their liking, we will talk to find an arrangement satisfactory to them to stay." "Mama you have gained a loving daughter," said Sam.
"Child, do not encourage me to make too many wishes."
"I will tell my parents that I have met an equally loving family and they will want to come to see you."
"Let me talk to Bosco." Sam moved and whispered to Bosco to approach Mama.
"You look cheerful, and I am sure you are ready for marriage," said Mama. Bosco smiled and said,
"I am certain, Mama."
"Have you talked to Sam's parents?"
"Yes, I saw them. They are very poor. They live in one room flat in a dilapidated suburb. I wish they could move to a better residence but no one will rent them because the man is disabled with no sure income."
"Bosco, you have many unused buildings, refurbish one and allow them to move in," suggested Mama.
Bosco hugged Mama and said,
"They will settle here before we are wedded."
Sam saw the hug and went closer to Mama. She responded to Mama's suggestion and said,
"I will make preparations for them to come as soon as possible."
"What are you waiting for?" Mama asked.

A Dot on the Timeline

Titti was checking that the process of every plan was as complete as possible. As the leader, she took full responsibility. Every night Titti checked and rechecked that no procedure had any glitches. Invitations to weddings can be a source of arguments. The bride may want many people but the groom may want a few or the other way round. The parents may want to invite many of their friends and that may cause some tension. Some sisters or brothers may want more invitations and they have to compromise. At Bosco and Sam's wedding, there were no problems, because there were only the parents of Sam and Titti. There were no brothers or sisters or any significant number of friends.

The invitation cards sent to all neighbours raised great curiosity. Many minds questioned how the old man could support a despicable enemy who had harmed many in the farming community. How the brutal man found a wife was a question no one could answer. The interesting question of why the old man used his house as a wedding venue was on the lips of many neighbours. Many neighbours accepted the invitations, not because they loved Bosco, or thought that Papa was acting wisely but to see whether what was happening was real. Many neighbouring farmers thought that Bosco was about to get his comeuppance. Some people thought that there would be very little food or booze to humiliate Bosco. Some thought that some kind of unsocial act against Bosco was going to happen.

Papa had insisted all the time that he would release the secret at the wedding. Some farmers expected something catastrophic to happen and for the bully to get what he deserves. The family kept their lips tight despite the numerous rumours.

Victor asked for patience to wait until the wedding day to reduce the barrage of questions he could give a response but did not want to answer. He promised to tell everyone at the wedding. The women's friends approached Titti to know the mystery. She said,

"In marriage, all secrets are laid bare. We say nothing until the day. You should not listen to rumours until the day. We want to make it monumental."

One friend of Titti said,

"I hope the old man's plans to get rid of the cruel man for good will succeed."

Sam did not know the wicked past of Bosco. Titti and Victor did all they could to shield Sam from rumours and innuendos of neighbours who hated Bosco.

Titti said to Sam,

"You might hear comments from neighbours. If Bosco has not told you, please ask me first before you believe because some farmers have done bad things and Bosco could be protecting his farm at times, not in the best of ways."

Titti had spoken holding crossed fingers behind her back because it was not the whole truth.

33

The busy schedule made the month appear a very short period to Bosco, Victor and Titti.
Victor said to Titti,
"Bosco appears deep in thought as he counts hours to his marriage."
Titti said,
You were not different on your day. Remember it is a mark on your life that you want to happen without a blemish."
Bosco overheard the chat and said,
"Nephew, you are correct. It was a tense moment. It is a big change for me from loneliness to a found family and in no time, I will have my own family. It is change, hard to take in quickly. "
Victor smiled and said,
"Uncle you are not alone but you are surrounded by family that is growing to become large."
An early morning rise on the day of the wedding witnessed a cloudless sky and a burst of bright sunshine. Bosco thought that was a good omen. He told his nephew his thoughts. Victor laughed and said,
"We are farmers; it is a good omen whether it rains shines or anything in between."
Bosco said,
"Nephew, let us agree that there is no bad omen, especially today."
The tick-and-tock sounds of the clocks continued as always.
The sheep, the pigs and a large number of chickens were in fear. Who would not be worried knowing the fate of many may occur on that day?
The days to rest and collect dust for the wines in the cellar were over. The bottles cleaned and chilled were ready to offer their

long-time stored contents, for the guests to consume with pleasure.

Papa and Mama sat dressed as crowned heads. Victor escorted Bosco to the table decorated with flowers and embroidery to wait for Sam. Titti ensured that Sam flanked by Alena and Sofia, showering her with flower petals, looked flawless in her dazzling dress as she walked slowly, along the carpeted passage to join Bosco. She walked slowly for all to see her and admire her attire. The appearance of Sam made all guests stand and gave Bosco a feeling of invincibility.

It was then and only then, that Bosco could be ready to say the immortal words, "I do."

Sam without hesitation uttered the same two words in front of the appointed commissioner.

The raucous applause of the guests marked a day in the time of lives of two people Sam and Bosco. To Papa and Mama, Titti and Victor and their children, it was a day to remember when the employee became the uncle's wife.

Papa and Mama expressed their feelings of happiness through smiles and greetings to every visitor. The visitors, however, wanted to know how Papa managed to give Bosco a wedding party, a person who was recently his enemy and to many of the farming community.

Bosco looked worried but with Victor beside him, he calmed down. To show to the crowd that Bosco had the respect of a son, Papa and Mama reassured the guests that all was well with Bosco. They did not need to tell Sam that she was a family member and a loving daughter because she knew that. They intended for that day to be memorable, a spot on the family timeline. Papa and Mama wanted to remember it as an important mark of peace and love.

The guests had sat for a while drinking good wines, beers and juices. Many were less tense and able to talk to those they did not previously know. In their exchange, one could hear the word speech. Giving a speech is a usual part of a wedding ceremony. In such speeches, there would be advice or guidance and at times, humour that could cause a burst of laughter. The guidance may prove valuable but the humour may tell a secret. At Bosco's wedding, someone had to let the secret out of the bag.

34

Victor stood up. Everyone knew that he was going to give a speech.

At that moment, one could hear a pin drop. No one wanted to miss the events that made Bosco change from a family enemy to gain the respect of sitting at a high table. Victor checked whether every guest had a full glass of wine. He did not rush, he wanted to extend the suspense time as many guests were waiting to hear how Papa tamed Bosco. They were keen to understand how Bosco happened to sit in Papa's house. Some thought about how a rude man like Bosco could find a beauty like Sam. Victor gulped some wine to soothe his throat and then he said,

"You may have questions that start with when, why, how or what. Many of you are wondering and curious to get answers to those questions. You do not have long to wait. The man you see in front of us, holding his wife protectively, looking humble and handsome is my uncle."

There was a slight murmur from the guests. Some people glanced at Victor and at Bosco to see any family resemblance or features that could suggest a family link. If anyone claimed to see the link in that dazzling set-up, was likely fibbing.

"One day, in anger, after an argument, I told Bosco of my wish that if my parents were alive they would strangle him to death. I went on to tell him that my grandfather was too frail to do that. Little did I know that my grandfather had a cutting of a newspaper dated about thirty years ago that reported the death of my father and mother."

More murmurs and sighs showed that the guests were sorry to hear the news. Some wondered how a newspaper cutting could solve a quarrel between two people.

A Dot on the Timeline

The speech told the full story of how they discovered Bosco whose name was Charles. Victor sat down and looked at Papa for approval before taking a sip of wine. All the guests filed in a single line to shake the hand of Victor's uncle, Charles Bosco and Sam, his wife. The mood of many gusts changed. Those who expected a showdown realised the ceremony was genuine.

The guests' chatting became raucous. They knew the man who terrorised them as the uncle to Papa's grandson. Some guests started to gloat that they could see a family resemblance between the grandson and his uncle. The guests who were mainly the neighbouring farmers, who had suffered Bosco's cruelty, toned down their hostility and started to enjoy the wedding celebrations. Bosco made more friends than foes as he sat next to Sam. They believed that Bosco would change from a hostile land grabber to a gentle happy, kind uncle and a happily married man. The revelation was like a fantasy story or a dream but it was a real situation. The story became a village gossip. At the same time, Victor's speech marked the start of a respectable life for Bosco. Bosco whispered some words to Sam. The smile made everyone look.

Titti stood up, straightened her dress, looked at all the guests, saw happy faces, smiled and said,

"To discover that a person you consider a long-time enemy is a long-lost relative, you will think you would instantly celebrate. This was not what happened in our case. There was a shock, there were tears and there were worries that the search could be wrong. Bosco had a set of bracelets from his mother. He gave it to each of my daughters. Mama was surprised to find that Victor's mother had the same bracelets as those from Bosco. The incident happened emotionally without any preparations. It removed all doubts that the search could be wrong. That reassured Victor that Bosco was his uncle. We are sure he is our dear uncle and a loved member of this family. That was the first surprise from Bosco. There followed a very loud clapping of hands as Titti paused to sip the wine.

"We did not know that at first Bosco had so much love. He had to share it with the most beautiful girl that has made the family a beacon as a lighthouse to lost ships. Their bond has strengthened the family and that is the second surprise Bosco has brought to this family. The third surprise was that until a few minutes ago, you were eyeing him as the despicable unsocial land grabber but you are shaking hands signifying friendship. You can see he cannot stop smiling. He is among friends. The surprises for the future are what I cannot tell you. I will let him tell you that story himself when he chooses to do that. It could be a fourth surprise."

Titti raised her wine glass and when all were standing, she said, "To our uncle Bosco and his bride Sam, we salute you."

The noise from clapping, ululating and talking, vibrated in the hall for some time preventing anyone from talking.

Papa had sat quietly listening. Titti's speech impressed him such that he had to speak and say something. He asked Mama to pass him his walking stick. He stamped his stick onto the marble floor to get the attention of all in the hall and then he said,

"I would like everyone here to forget the past and give respect and love to Bosco as my son. In return, he will respect and give support to all."

The rigorous clapping of hands indicated approval and sheer happiness. The thought of peace in the farming valley encouraged many to celebrate the end of hostilities. The number of empty wine bottles was a sign that there was genuine enjoyment but there was a hint that many had surpassed their consumption limits. It did not matter; they were having a good time.

Mama wanted to express her happiness too. She stood, holding Papa's shoulder as a prop. She waited until everyone was quiet and then she said,

"When I saw Sam, I loved her instantly and wished for her to stay and remain here forever. The miracle happened. You can see she is here happy, protected and loved. You can trust Mama to make a wish for you."

A raucous laughter erupted with some guests shouting repeatedly,

"Trust Mama's wish"

Mama continued to stand, and the crowd calmed down prepared to listen as she said,

"When Bosco came to my house with Victor calling him uncle, I was mortified. Your science today tells you that you can genetically check for family links. I used my science. It is a careful observation of family traits. I saw the smiles. I concluded Bosco was my grandson's uncle, as good as the son we lost. You can see now, he has made all of us very happy and contented."

Mama sat down as the guests looked at Bosco and Victor to see what Mama said and made their conclusions that no one dared to share.

Papa whispered to Mama and said,

"I could see the trait too, you have openly declared it. You will get feedback that will prove your observations as correct."

The guests continued to enjoy the generously supplied food and booze. The courage to speak to congratulate the couple was gaining momentum as Titti's wine was starting to show its effects. Some gave small magic shows and some started impromptu singing.

Alena and Sofia started dancing to the background music. They were bored of talks by adults. The adults who were bored of speeches started to dance to encourage the little girls who showed their dancing styles.

Papa noted Goodwood and Bosco shaking hands. He elbowed Mama to see. They smiled with happiness and started bragging about their achievement of bringing peace to the valley of the farming community.

The guests were still sitting, chatting, laughing and at times looking seriously while exchanging some words of wisdom and were not showing any hint of leaving. However, Bosco and Sam had planned to travel to the city. Victor imagined how his uncle who had not been in the city would enjoy the lights, sounds and sceneries. He wished them a blissful future as they started their journeys measured in distance to the city while guided by time through their marriage.
Victor called for attention and said,
"Please, lift your glasses to wish the couple a safe journey."
The guests said something mollifying as a farewell. There were no enemies in that hall. A few had taken the alcoholic beverage above their limits and did not care whether he was a friend or foe. Those grunted inaudibly and that did not cause any problems. The wedding celebration, the festive moment, the food or the booze could have wiped the slate clean of all the profanities uttered in the past.
Papa revealed to Victor his thoughts on the family situation:

Time has strange abilities to hide or reveal the past
It can smoothen or roughen the path to the future
A dot on the timeline only scripts the present moment

Victor said,
"We have learned to cherish all things offered, or those denied by time."

35

Victor was worried. He clasped both hands of Titti and said, "I hope and wish that all goes well with them. I am used to having an uncle."

You could think he was behaving stupidly worrying about his uncle going away. It was a journey like no other. Bosco was nearly fifty and it was his first vacation as an adult. He was not alone. He was going to a city far away with someone else he had to care for and protect. It was like two journeys combined, a movement and a new experience. Victor's fear was whether Bosco could navigate through the rigours of sharing and living with someone else after the long period of a lone existence.

It was late afternoon. The car decorated with flowers carrying Bosco and Sam was about to depart. There was clapping, ululation and words of encouragement shouted by the well-wishers. The beauty of the colours that decorated the car, made it unmistakable to bystanders that it was carrying a bride and groom.

Sitting in the back seat of the car, the newlyweds were silent; not because they wanted quietness but because they were keen on listening and looking at all that was taking place outside their car.

"It is time to go," Victor told the driver.

They hardly saw the outside through the windows due to the confetti of flower petals that showered down past the windows as they drove away slowly through the corridor of friends and family. Sam and Bosco appeared as dignitaries proud of the treatment. Bosco looked at Sam, held her close and kissed her. There were no words to describe the mood but a touch was sufficient. The car left. Papa turned to Mama and said,

"They were going on a break that both had not experienced. The city visit, the bright lights, the entertainment, the speedy lifestyle

and the start of their love life were matters bound to be overwhelming. I hope Bosco will cope."
"He is in the hands of a strong woman, he will manage," said Mama.
"He will have time to reflect on his past and the event of finding a nephew may sink in," said Papa.
Slowly, the guests showed that they were ready to leave. They shook hands with Papa and Mama with words of praise. Victor, Titti and the children went to hide inside the house to avoid the long chats meant for goodbye until all the guests had gone. It was late as the family was able to meet and relax. The family realised some home truths.
Victor had to manage two large farms for at least a fortnight. Mama was in thought,

It is not long
Eight would be the family size
A wish for two children
A hope for possibly more

She shared that thought with Papa by whispering. The two, holding hands laughed as they strolled to their bedroom. Victor and Titti looked at the happy grandparents and Titti commented, "That laughter is a sign of approval for our hard work in making the ceremony a success."
"I am tired," said Victor as he dragged his feet walking to his bedroom. The thoughts about the journey of his uncle persisted in his mind ceaselessly.

36

The pleasing thoughts of Sam as expressed in her smiles were good enough to make Bosco relax despite approaching the town of Moshiana, a town he dreaded.

The road to the city passed through Moshiana. It was a small town made black by smoke and coal dust. It was not dark for lack of lights but covered in black dust. It was the centre of coal mining and the coal dust had coated every part of any erect structure including the surface of the roads and road signs. The streetlights had too much soot that blocked the light. The dark streets became a haven for criminals who roamed undetected, making crime rife. The driver had to slow down. It was dangerous to drive fast, as he could not read the road signs. The slow-moving car was in danger of attack by car thieves. The road markings and the pavement had layers of coal dust, forcing them to stop and at times seek assistance to find the road out of that cursed town.

Bosco had visited that town as a reckless young man. He did some things he wished to erase from his memory, a past he did not want his wife to know because he learned nothing but the ugly side of life. He did not want to be there, remember his ordeal or tell his wife. He wished there was another route to the city without passing through Moshiana.

His wife and her parents lived in the suburbs of that town. He was glad they had moved to live in a farmhouse in his village. It is not good to keep too many secrets from a wife; Bosco thought in fear his wife might know something that could point the finger at him. He relaxed as he saw the sign directing the road to the city. He breathed a sigh of relief and then said to Sam,

"I worked in Moshiana as a youth. It was not as dirty as it looks now."

Sam said,

"It will not change as long as it remains a coal mining town. I lived and worked in the outskirts of the town, it was not as dirty as the centre of the town. Thanks for giving my parents a better home."

Bosco smiled as they were past Moshiana and approaching the glittering lights of the city. Bosco was proud and imagined what he would tell his nephew on return. Although he was wiser, richer and not alone, Bosco's experience of urban areas was always frightening despite having Sam by his side. The wide and straight roads enabled the driver to speed up. The bright lights, the high building structures and a variety of new noises signalled their arrival at their destination in the city. They were tired and needed a rest.

A variety of noises mostly of engines and exhausts woke Bosco and Sam early in the morning on the first day of their short break. That change of morning noises for a moment was a fright but soon, they relaxed in knowing there were neither pigs nor sheep where they were. Bosco and Sam planned their visits to theatres, cinema halls, art galleries, zoos and botanical gardens. On each day, the plans went on without any adverse incidents. They were all smiles Bosco was impressed with the array of flowers in the botanical gardens. He said to Sam,

"We have to plant some flowers in front of our house to make it beautiful."

"I thought you did not like flowers," said Sam.

"I want to surprise Papa, Mama and Victor," said Bosco.

"We have to build a fence to prevent the sheep from grazing on the flowers," said Sam.

"I will do that dear," said Bosco, as he laughed proud of his suggestion. Bosco looked at Sam smiling and said,

"Did you enjoy the visit to the zoo?"

"Yes, I did, the animals were well cared for."

"Would you know if they were sick?" Bosco asked.
"Yes of course," said Sam.
"How can you know? Animals can neither talk nor show signs," asked Bosco.
"I have worked with you for more than a year. Are you testing me now?"
"No dear, those are wild animals. They could be different from domesticated ones," said Bosco.
"We train to see the signs. You are untrained; you will not see the symptoms," said Sam smiling. Bosco detected the possibility of losing the argument and went on to talk about plans to improve his farm conditions to match the zoo's clean surroundings. They chatted until it was time to go to sleep.
Seven days of rest were enough for Bosco. They talked about everything, then there was nothing to discuss. Bosco started to worry about the state of his farm and convinced Sam to return two days earlier. Bosco had developed the confidence and courage to think that nothing could ever go wrong again. He was ready to face anything the world could dish to him.
Sam said to Bosco,
"We will pass and stop for a day at Moshiana to say goodbye to my friends."
"Sure, I will be happy to see your friends."

Bosco was barely awake when they arrived at Moshiana. It was not at night but at midday. He smiled, putting on a brave face to overcome his fear of Moshiana. Sam's friend, Harriet, gave Bosco a good reception and that helped to calm Bosco's nerves. He was relaxed and courageous to suggest a trip to the town. After ten years, he thought, he would not know anyone and no one would remember or recognise him. He followed Harriet to the betting shop where she worked. The shop was still open. Work colleagues paused to see Sam and admire her wedding ring. The magnificent reception kept Sam smiling. Harriet had replaced Sam as a manager of the betting shop, a transfer and promotion from a smaller shop.

Bosco walked to places that Sam and her friends used to frequent. They walked past the bank where a robbery took place leaving a man dead. Bosco knew of that robbery but said nothing. He saw the huge boulder of granite stone. He remembered what he did there but could not tell.

Worrying Times

37

Bosco found the dinner scrumptious He suggested to Harriet and Sam to check out Moshiana's nightlife. A pub visit was the choice they made. Bosco flanked by two glamorous lovely women was all in smiles thinking he was the envy of other pub customers. Bosco sat in a corner and looked at the people but did not appear to recognise anyone in particular. He was not good at remembering faces or names. At the other corner were three people shabbily dressed and unshaven were drinking from almost empty beer mugs. They regularly glanced in the direction where Bosco was sitting. It was easy for Bosco to think that the men were glancing at the girls. A very old man, unwashed and unshaven wearing torn clothes sat beside the three men at the corner. He was alone not in the company of the three men. He appeared sleepy or drunk not bothering anyone. The torn clothes reminded Bosco of someone he knew when he lived in Moshiana but he did not utter such thoughts to Harriet or Sam. He was comfortable as he sat listening to Sam and Harriet talking about the goings-on since they parted while enjoying a glass of wine. Bosco noted that the glances from the three men at the corner were becoming more frequent. He was feeling uneasy. The more he looked at the people, the harder it was for him to block his bitter memory of his time in that cursed town of Moshiana.
A trip to the toilet brought many memories back when a man stopped in front of him and said,
"Hallo, Bull, where have you been? Do you remember me?" His white hair was unwashed. Coal dust covered it as well as his worn and torn clothes including his skin. His toes protruded outside the old leather boots that were coal-coated. The need for money was probably what Bosco thought but the name Bull, the school days' nickname, made Bosco pose to ask,

A Dot on the Timeline

"Who are you?"

"You are in danger, so you better listen very carefully," said the old man.

"My name is Timson." He did not recall that name but for anyone to call him Bull, they certainly knew him. Bosco used the name Bull, his name as a bully in school and again in Moshiana when his money had run out and was in the town unsure of the next meal.

Bosco's eyes opened wide as he listened.

"Ten years ago you were sitting in a getaway car. I was sitting on the boulder where you also sat many times before that day, recording the goings-on of the bank. You will remember me because you used to buy me a drink whenever we met. You were kind to me and I liked you." Bosco remembered the incident and moved closer to Timson despite his repulsive body odour originating from his unwashed body, clothing and the bad smell of his breath. Timson continued knowing he had Bosco's full attention saying,

"As I sat on the boulder watching the bank, I heard a gun sound from inside the bank. The robbers, your colleagues, could have shot someone. I did not want you to hang so I called you out of the car, and dragged you away to hide under my blanket behind the boulder. Soon after we saw your friends coming out with sacks of money but none could drive. We hid behind the rock. I covered you with my blanket when the police arrived and arrested all as they fiddled in the car. I went to court to listen to the case to tell you but I never saw you again. They received a twenty-year sentence. They would hang but for one legal glitch. A ricocheted bullet killed the man. They came out of prison after ten years for good behaviour and are sitting near me in the corner. I do not think you recognised them. You have a very bad memory. Bosco's heart was hammering. One of them may have recognised you but I do not think so because they would have pounced on you. I heard them arguing about which girl sitting with you was prettier. I think I deserve a reward."

Bosco knew he was in danger. The fast-beating heart and sweaty hands meant he should leave fast. He remembered the events. He remembered the old man's actions to save him but not his name. He put his sweaty hands in his pocket and gave the old man some money. Timson stretched his hand again. The gesture meant that the amount of money he gave was not sufficient. After two such attempts, Timson nodded in agreement and said, "You should not stay here for long; you have to leave promptly because I heard one of them saying he would cut all your fingers to match the years of detention. The other man said you would have to pay all the money they lost."

"Thank you, Timson, I remember," said Bosco.

Bosco had to think fast. He had tried to forget the memories of that town but that town never forgot him. After pronouncing some profanities about that town, Bosco eventually said, "Timson, do me a favour. Tell the two women to come out. Let them leave slowly without causing an alarm. Tell them that I am out getting a taxi as I feel unwell."

"Do you live in another town?" Timson asked.

Yes, in a village. I own a farm in Riverside Farming Valley," replied Bosco.

"When I get enough to pay the fare I will visit you," said Timson.

"You are welcome," said Bosco, encouraging Timson to hurry up to inform the women to go outside to meet him.

Calmly, Timson returned and passed to tell the two girls to leave. He took Bosco's pint filled it with drinks from the girls' glasses and behaved as a bar worker collecting glasses. He often collected glasses from tables to get a free drink. That clever move did not cause any alarm. Timson calmly returned to his seat drinking a pint of beer flavoured by the girls' wine.

The convicts at the corner asked,

"Where did you get money to buy a drink?" Timson stood in front to block their view to ensure they did not notice that the two girls and the man had left and then he said,
"I robbed one rich drunk man. He dropped his wallet when he took out his handkerchief to wipe his wet hands."
"Where is the wallet?" asked one of the robbers.
"Have you lost your cunning brain? Do you keep wallets or do you keep the contents?" asked Timson.
"OK, clever clog, get us a drink," said one man.
"When did you buy me a drink?" Timson asked.
"One day, they will catch you robbing, who will help you selfish bum," said the other man.
"I will not expect help from a robber. You never do anyone anything good," said Timson.
"We can rob that cash from you now," said the other robber.
"When did you start robbing old men?" asked Timson.
"Don't be insulting to an oldie," said the other robber.
"Ok, but when we meet again you will buy a drink for me," said Timson as he gave some money to one of the robbers to go to the bar to fetch the drinks.
He had the feeling of exerting power as he sat next to the robbers and talked as equals to them. Timson had for a few minutes tasted the power money can give to a person.
They believed him and after enjoying the pint, they realised that the man who was sitting with two women had gone. They were happy about the drink and did not spare a thought or care about the man with two women.
Bosco was silent. He sat in the transporter while Sam and Harriet walked slowly to the vehicle without showing any need for urgency to leave. They were oblivious to the troubles of Bosco. He asked the driver of the taxi to move fast. Bosco did not want to face the ugly dirty town he considered unfriendly. After ten years, there was no change in how the town treated him. He could not express his thoughts aloud. He was certain that the

hurried exit from the pub would need an explanation. He was then a married man. He could not hide such incidents. He sat calmly but heavy thoughts weighed his mind. He looked at his wife; she appeared relaxed as she talked to Harriet.

Bosco arrived at Harriet's house. He was planning a quick getaway to his farm. He planned an early morning drive. Sam agreed to Bosco's rushed plans without hesitation. Sam had noticed some anomalies. The gloom on Bosco's face, the hurried exit from the pub and the old man, the informant, were the glitches that perturbed Sam. She planned to ask when the situation had calmed down so she reserved questions for another time when at home. She was then a wife; she had ample time to get out the truth. However, she did not know the difficulty of getting the truth that no one wants to tell. How love and trust could confuse judging a misleading human behaviour. Bosco was imagining what his wife was thinking. He could not tell his wife a lie, it was too hard, and he kept silent. To keep in mind such secrets for long is very troubling. To get peace of mind, he thought of telling his nephew.

38

You will not usually quit abruptly from a pub unless there is a serious problem. A pint of beer half of it consumed and left on the table was a sure sign there was an underlying emergency. Bosco did that when he knew he was in danger. His survival and that of the two women was the reason for the action. The quick departure was bound to be a source of many questions. He was busy preparing a lie to tell. He could not tell the whole truth at that moment or possibly at any other time. However, he had to say something. Before he uttered a word of his perfect lie, he noted a picture on the table and looked at it. He stood up, went closer, picked it up and asked,
"Who is he?" Harriet looked at him, her face was sad when she said,
"It was my father, why are you asking?
Bosco said,
"The face looks familiar to someone I have met before."
"I do not think so because he is not alive. Someone killed him during a bank robbery ten years ago. The police caught the robbers. They got a twenty-year jail sentence but were out after just ten years, for what they called *'good behaviour'*. Can good behaviour bring back my dad?"
Harriet started crying. She calmed down after a while as Sam embraced her. She was able to talk again and said,
"He had gone to arrange payment for his miners when in broad daylight the robbers got into the bank to rob. Why did they have to shoot him? He was just a customer. The men at the corner in the pub seem to glance at me and I strongly think they were the robbers I saw them in court those ten years ago. They were younger, clean-shaven and now older and rough looking. It is

hard to be sure. But I know they are out of jail and back to society for good behaviour."

"Good behaviour my backside, they pretended to be nice to get that jail time reduction," said Sam annoyed with the legal system. "It is justice they said. What justice did my dad get?" said Harriet starting to cry again.

Bosco's legs wobbled. The sweat from his armpits was visible. He sat down, wiped his facial sweat and without looking at Harriet, he said,

"Sorry, for the tragedy."

"No, it was a cold-blooded murder for which the punishment should have been hanging by the neck until they died. It makes my blood boil to see them laughing as they enjoyed their drinks. When I prove their identity, I will give them the treatment they deserve for killing my father."

"What will you do Harriet?" asked Sam.

"I will get the services of some hoodlums. They are cheap to hire. A small sum of money will be enough to wipe the smile off the faces of the killers by knocking out a few teeth," said Harriet with her mouth twisted.

"Please Harriet, do not do that. The thugs you use will demand more favours, especially when they run out of money. You may end up a victim. There is no honour among Moshiana's gangsters," said Sam.

Bosco was tongue-tied. He was confused. He was one of the goons. The bitter words of Harriet forced Bosco to concentrate on looking at the man in the picture to hide his emotions. Sam noticed the continued look at the picture and said,

"If he resembled someone you met, he must have made a good impression on you because you have not taken your eyes off that picture. I know you are not good with names or faces but you will remember eventually."

"I think he was the man who gave me a job," answered Bosco.

"What job?" Sam asked.

"As a driver in a security firm he owned, but I am not entirely sure as I met him only once."
"Do you remember his name?" asked Sam.
"No, I am bad at remembering names as we only referred to him as the boss," said Bosco.
Sam looked at Bosco and noticed the onset of his gloomy face and said,
"We better leave this town and go home now. This town appears to make you sad."
"Yes, it is a town that brings misery to people. We shall leave tonight," said Bosco. Harriet heard the talk and did not want them to leave saying,
"We have a cup of tea and a cake I made earlier. We will calm down and prepare to leave early tomorrow morning. I am too distraught. Please do not leave me in this state. I shall not cry again. You can leave in the morning." She thought they were leaving due to her sobbing.
Sam said,
"We will have lots of time to chat. You are welcome to our home, in the Riverside Valley farming village any time you choose to come."
The invitation made Harriet soften her stance and allowed Sam and her husband to leave after the tea and cake. It was late in the night but Sam and Bosco had to leave Moshiana. The journey homeward appeared to take a long time. Bosco could not sleep. He watched the changes in the vegetation, as it was a moonlit night. That distraction did not remove from his mind the past and bitter thoughts of his life in Moshiana. Timson saved him with his intervention. He smiled to note that Timson did a good turn, just to repay a pint of beer that he had bought ten years ago. He relaxed and dozed off once he was a distance away from Moshiana.

Sam nudged him to say they had arrived home. It was in the early hours of the morning. Bosco was very quiet when they arrived. The resemblance of the facial features of Harriet's father, the man in the picture, to those of Bosco's adoptive father, weighed on his mind. He did not disclose that thought to Sam. He was afraid to lie and feared telling the truth. Keeping that secret and guilt was probably what made Bosco less jovial when he arrived home. He also knew that secrets never remain secrets forever as time has the power to reveal. He smiled as he approached the family members. On the veranda, he opened his bag and had a present for each member of his family. He was proud as the little girls received him with hugs and kisses.

Victor was happy as uncle Bosco returned. The welcome, the reunion and the breakfast reminded Bosco of the privilege of having a family. That great breakfast with the eight members of the family pleased Papa. With Bosco and Sam, there was a chance that the family could grow. Papa thought how the moment we describe as now or present time is very brief. In a flash it becomes the past. He did not want the future to be as brief as the moment we call now. He asked that all the evening meals and breakfasts had to be at Papa's table, an idea that received a nod and a smile without exception.

The good news of increased sales of wool, livestock and wine added to Bosco's excitement and for a brief period, he forgot about Moshiana.

Papa's revelation that he had been resident in the city started a conversation with Sam who told him of all the places they visited. Papa knew many of the tourist sites and the exchange between Papa and Sam became the main talk at the table.

Some changes in the farming community had occurred without the awareness of the family. Victor noticed that the neighbouring farmers were coming to Papa's house one after another. He thought they were curious and wanted to know about the city experiences, to see the couple or to hear the travel gossip. The

A Dot on the Timeline

number of visitors increased as some visitors remembered the taste of the wine and were hoping for an offer of a tumbler or two of the wine.

It was not long before Victor realised it was like a gathering of friends and asked Titti to supply some bottles of wine. Goodwood arrived showing off his milk and cheese packaged for sale by offering it as a present to the couple. The cheese and milk present reminded Bosco of the plan he ditched. Goodwood however, revealed more good news saying,

"There is a plan to build a bank in the village. It will transform our village into a township and farming trade may increase."

Bosco said,

"It is a good idea but we must start a group to provide security to ensure we do not become victims of hoodlums like those in that town of Moshiana." Goodwood was happy to find some common point of agreement with Bosco and then he said,

"We act on it as soon as the bank plans are in progress."

39

The neighbours had Bosco in their good books as a respect to Papa. He did not deserve any good deed because he had done nothing good to deserve another. Bosco was pleased to see the increased growth of business on his farm and the warmth of reception from neighbours.

Nothing is as rewarding as a bumper harvesting to a farmer. It is like a good turn, a reward from nature.

Victor had to find extra storage for his grain. His hard work was financially pleasing when he found buyers. Titti's winery was full waiting for buyers to empty it.

The flock of sheep had increased in size such that Bosco was about to run out of hay. That problem ended when Victor arranged for a city buyer.

Life for the family was leaning towards a healthy, wealthy and contented future. The bumper harvest, the happiness in the family and the great profits from sales was like a special favour from the heavens for the family. What else could anyone ask? Victor thought as he relaxed with his family around him. Bosco's contented smile said it all. The announcement by Sam that she was expecting was as if the skies had opened showering the family with confetti of gold.

It was after breakfast, that the whole family became silent and calm thinking about the good news while enjoying a drink. They had wished but did not expect that soon, the addition to the family. They became excited and joked with Sam and Bosco whether they were ready for a new title, a call from a young voice pronouncing mum and dad. Papa and Mama looked at Bosco and Sam smiling and then Mama asked,

"Do you know what a baby's cry signifies?" Bosco looked at Sam for an answer but Sam remained silent. Mama touched Sam's hand and said,
"The cry shows there is life, a sign of a family that lives forever."
Sam cried with happiness as she embraced Mama.
Victor said to Bosco,
"When you are a dad you will never be alone again. Your lonely days were over when you met Sam but now you are certain they are over forever."
Bosco stood up. He intended to make a speech to express his happiness. He glanced at the members of his family and at that instant; a lone figure appeared at the gate and interrupted the start of the speech. The whole family turned to face the stranger at the gate. Victor beckoned him to approach. The man came closer and Bosco recognised him as Timson. He was not a stranger to him. He was as dirty as the last time they met. In an elated mood, Bosco invited Timson and gave him food. He knew he would be very hungry. All the other members of the family looked on. Their brains were churning many questions to ask Bosco. They knew Bosco was kind but that behaviour towards a vagrant was above any expectations. Bosco did all that with Timson without saying a word to the other members of the family.
He was happy to share his food with a stranger who in Moshiana was a protector. After a good feed, the stranger said,
"My name is Timson. Thank you for the food. I will do some jobs to pay for it, as I have no money at all." Papa and Mama looked at him with sympathy and Papa admired his polite nature by saying,
"We see you are a friend of our son, you are our friend, relax and you are welcome." Bosco had to keep his friendship with Timson who saved him from a possible attack and concealed a secret that Bosco did not wish to reveal.

Mama, Titti and Sam were too busy thinking about the news that Sam had made public. Papa, Victor and Bosco planned what to do with Timson. The family was unsure whether to invite Timson for the evening meal or not because of the filth. Bosco offered to take Timson to his house, a gesture Papa and Mama found impressive. Victor followed the events and waited patiently for a suitable moment to ask his uncle about his new friend. Bosco and Timson returned to Papa's house. There were smiles. Timson had showered, shaved and had a new set of clean clothes and new boots. The family looked at him and could sit with him to enjoy an evening meal. With new clothes and brushed teeth, he looked like a young man although he needed a walking stick like Papa. Timson smiled a lot at the sight of a table with good food in abundance. It was like a dream to him. He had forgotten how people sit and eat their dinners. He had forgotten table manners and the use of the usual cutlery. He looked and copied Bosco as they sat to eat. He was a vagrant but not a fool. He did not cause any discomfort at the table.

40

Mama and Papa were sure that Bosco and Timson knew each other before that moment and they wanted some answers to their curiosity. She knew how to get answers. She thought of starting with the stranger. It was after the evening meal. She gave Timson a glass or two of wine and then she asked,
"How did you two meet?"
Timson stood up, and in a humble polite voice he said,
"Your son was very kind to me when he was in the town more than ten years ago. He had money and he was generous to me. I carried coal to houses and he drove a car carrying security officer for the coal miners. I gave him a bit more coal in his order as he always bought me a beer at the pub where I stayed especially when it was cold. I was and I am homeless." Bosco was listening. He clasped his hands tight in fear that Timson would tell the whole truth. He breathed a sigh of relief when he heard Timson saying,
"Suddenly, I did not see him until last season when I recognised him and he did what he always did, buying me a drink but this time invited me to his home. I am very proud to meet his great family."
Mama noticed the smile after the reply and she looked at Titti. Titti nodded. The two women had seen something that needed a bit of thought.
Bosco recovered from his fear and said,
"He will stay in one of the houses. I will ask him to watch and ensure that the sheep do not stray and he will earn his keep until he returns to Moshiana if he desires."
"No," Mama said, "he will stay in our spare room. He is an old man we cannot leave him to stay alone." Papa agreed. Timson

smiled, he had food, a job and a room to stay. He could not find his voice, but he nodded to say thank you.

Titti and Mama were talking and Mama said,

"The forging of friendship for two people with such a great age difference is unusual but possible. There must be something more than that pint of beer that made Bosco and Timson friends."

Titti replied and said,

"Bosco must have tried to get friends and helping the old man could be the start."

Mama looked at Titti and said,

"Bosco was a loner but Timson appeared to have started the friendship. Something happened to bring them together."

"We will find out in time," said Titti.

Mama smiled and said,

"Did you see the teeth?"

"Yes, Mama that was a surprising similarity."

Mama smiled again and asked Titti,

"Have you noted any other similar features of Bosco and Timson?"

Titti looked at Mama and said,

"The voice was too close in similarity to that of Bosco. You cannot always tell who was speaking."

"Yes my child, that is what is odd with this stranger that needs investigation," said Mama.

The laughter of the two women occurred in unison. It was an agreement, they would try to search Timson's past.

The delicate nature of such searches needed thought, Mama pondered.

Mama discussed the matter of Timson with Papa who confirmed his unease with the friendship between Bosco and Timson and wanted to know what they know about each other. Papa suggested starting with Bosco. He could not lie, Papa thought.

A Dot on the Timeline

After breakfast, Papa called Bosco when everyone had gone to his or her work and said,
"My son, you know you are family without a doubt. Please tell me how much you know about Timson or how much he knows about you."
Bosco knelt in front of Papa's bed and said,
"Pardon me, Papa, I have made some past mistakes but they will never be repeated. When I was living in Moshiana, I fell into bad company. Timson may know or may not know. He used to live rough and I took pity on him. I met him when he came to sell coal to me. When he begged for a beer at the pub, I gave him and that started a friendship. I forget faces but he never forgets. Ten years ago when I was financially desperate, I met three men. They suggested a very bad plan for getting rich quickly. It was to rob a bank. I was the driver of the getaway car." Bosco went on to tell Papa about the robbery and the part Timson played to save him from the gallows. Bosco went on to tell Papa about the incident that occurred at Moshiana when Timson saved him again. This concern by Timson, saving me from recognition by the robbers in the pub, strengthened our friendship and that resulted in the invitation.

Papa smiled, it was the truth. Mama nodded in agreement, patted Bosco's shoulder, and said,
"You are truly a family member; teach all the family to respect honesty. We will hear Timson's version. All you have said to me and Papa remains here. No one will know." Bosco, for the first time, had told someone the truth, the whole truth without hesitation. He was sweating, as it was a momentous task.
Bosco stood up pleased that his good turn to Timson got Papa's approval. He remembered Timson's clever account that did not tell all which saved him from embarrassment. He walked slowly to his farm with a spring in his step that Victor and Titti noticed. He smiled expecting a barrage of questions but there were none

155

from either Titti or Victor. Victor and Titti thought the meeting of Bosco with Papa and Mama went well because Bosco emerged with a smile and walked with a spring in his steps.

41

It was Mama's turn to get information from Timson. Mama had realised that Timson lived through a rough life but he had proved he was not a fool. Mama had to be tactful to get the truth and not a plausible rhetorical story.
After the evening meal, all the family members including Timson were enjoying a glass of wine before going to sleep.
"Timson," Mama called and said,
"You said you wanted to work. You will work but be treated as a friend and not as an employee because of your age."
"That is the best thing I have heard in my life, Thank you, ma'am," said Timson.
"You can call me Mama," Timson had a glass or two of wine when Mama asked,
"Can you tell us a little bit about your life?"
"There is not much to tell because I am a homeless person," said Timson.
"I was born in Moshiana. The dad was a coal miner and the mother was a seller of coal. They had two children, my sister and me. We helped our mother to sell coal. When I met your son, I was still selling coal although my father, mother and sister had died."
Titti was at a distance but when Mama's facial expression did not show happiness Titti moved closer and listened to what Timson was saying.
Mama whispered to Titti and said,
"Timson's family died when he was just an adult. The most tragic was the death of his younger sister.
"How did she die?"
Mama could not resist asking to know the cause of the death of the young girl.

"Please tell us how or what could kill that young person?"
Timson was close to tears when he said,
"She had a baby but that was not what killed her. It was the stigma of the unknown father. The looks, the talk and the innuendos added to her trauma. As a young woman and alone that was too much to bear. She probably died of loneliness and sadness. I had left home at that time and the parents were not alive. To avoid the shame, she left town for the village where no one knew her. When she was ill, she sent for me but she died and I could not afford to go to bury her. I stole some money from a drunken miner and went to put a tombstone on her grave and wrote on it her name."
"Did you find out about her child, your niece?" Mama asked.
The flow of tears was visible as Timson said,
"No one knew anything about the child. I do not know whether she lived or died. On my way back to the town on the day I wrote her name on the tombstone, I was thinking about ending my life too."
Mama looked at Titti. The honesty shown about the life story made them look at Timson with teary eyes. That part of the story of a young woman with a child and dying young made Mama and Titti look at each other with a mortal fear that there were some matching similarities with the story of Victor's grandmother. The looks, the voices, the teeth and the automatic friendship with Bosco, including the protective attitude that Timson showed, encouraged Mama to tell Titti,
"There are too many signs." Titti looked at Timson as he wiped the tears and said,
"Please can you tell us the name of your sister?"
"Are you trying to be rude? In my family, there is a custom forbidding mentioning of the name of the departed. It causes sadness," said Timson.
"No, we will not be rude to our son's friend," said Mama.

"I become sad whenever anyone calls that name. Calling her name will not bring her back. I think you are making fun of my poor existence. Don't you understand that my life has been nothing else but suffering?" said Timson moving his hands frantically to find his walking stick. Mama thought he could be violent and use his walking stick as a weapon. To calm the situation, Mama said,
"You are our friend, I meant it. You can tell us the name when you feel able to talk again."
Timson was shaky as he found his walking stick and stood up. He was probably very angry, as he needed extra support by holding the side of the table. He probably did not hear Mama's peace-making statement when he said,
"I think you want me to go. When people ask too many questions, you get the idea that you are not wanted. I have lived rough. It is not easy but I would rather live poor than allow judgement and insults by rich folk like you. I will go and will never come back."
Bosco and Victor were at a distance unable to hear the conversation but were able to see Mama, Timson in agitation and Titti as they talked.
Titti stood up rapidly held Timson's hands and said,
"We are not judging you or asking you to go. We want to know you and your family. A lot has happened in our family and we cannot ignore anyone, rich or poor, dead or alive. Please, we may or may not be able to explain why we asked for the name but do not go." Bosco and Victor noticed the commotion. It was not easy to explain the sight where Timson was standing and shaking; Titti was standing and holding Timson's hands and Mama looking on with fear. Bosco and Victor looked on but did not approach Titti and Timson. No one said anything at that stage. In silence, Bosco and Victor stood, looked and tried to listen as Titti comforted Timson.

Timson thought for a while, as Titti's tender warm touch hands showed friendship. Completely unexpected, Timson said, "Sofia was the name of my lovely sister."
Papa looked up when Timson uttered the name Sofia. Mama looked at Titti. That name was very important. Timson noted the stunning look and asked,
"Why are you speechless?
Titti replied saying,
My second daughter is named Sofia; would you rather stay to hear why we gave her that name?" Timson sat down.
Mama said,
"Would you want to continue in the morning or shall we tell you now?"
"I am too sad to talk now; can we leave this matter till morning please?" Timson requested.
"We will be calmer in the morning and you need a good sleep," said Mama as she stood slowly to guide Timson towards his room.
Before Mama spoke, Victor went closer to Titti and whispered, "Why are you talking about Sofia?"
"Shush, we will talk later," said Titti in a whisper. Bosco asked Titti in a whisper,
"I heard the name Sofia mentioned."
"Shush," said Victor with his hand on Bosco's mouth.
To prevent Timson from hearing the questions, Mama called and gave an order to Victor saying,
"Escort Timson to his bedroom and ensure that he is relaxed or asleep. Cover him properly before leaving the room."
Without knowing the reasons for that order, Victor obliged. Timson used his walking stick and needed Victor as a prop as he was feeling very weak and in danger of falling. He took some time to reach his bedroom. On his bed, he said to Victor,
"You can go away now; I am safe here as I will ever be. Remember I used to sleep rough outside on cardboard with little

covering. This room with a bed and warm coverings is paradise to me."

It is possible that in his bed, Timson could think of Sofia and lamented:

> *"I cannot sleep guilty due to your suffering*
> *My failure to help you may have caused your death*
> *You died alone, calling me and I let you down*
> *The unpleasant memory has made me ill*
> *I am sorry*
> *Alone you died and as a vagrant, alone I will die."*

The name Sofia and the link with Timson's sister were at the forefront in the minds of Victor, Titti, Bosco, Mama and Papa as they woke up in the morning.

42

Mama was tense and in deep thought. The name Sofia was a matter of further investigation. Timson was in stress and that was their concern. She checked whether Papa had any new ideas by saying,
"A memory of the past shapes future ideas." Papa was standing by the window watching Victor who appeared to be in deep thought.
"Please do not test me with your puzzles, this early hour of the morning, I am busy," said Papa. Victor was standing by his window breathing in the morning fresh air probably thinking about Sofia.
He was brushing his beard with his fingers. Papa opened his window and disturbed him.
"Good morning Papa," Victor reacted.
"What are you thinking about?" asked Papa.
Before he could answer, Mama went to the window and said to Victor,
"A memory of the past shapes future ideas. Do you know what that means?"
"That is a question suitable for wise men and women but not the youth like me," said Victor.
Papa laughed and said,
"Now you see, you get your match."
"I was trying to know how to talk to Timson to make him talk to us openly," said Mama looking at Papa seriously.
"I heard the commotion last night. I think he thought it was an interrogation. He is our guest. The questions were too soon," said Papa. Mama shook her head and said,
"I wish I had asked you before we started."
"Victor," Papa called and said,

"Are you making a wish?"
Victor smiled and said,
"You read my mind because I was making a wish about my family."
"Wishes may thrill, may disappoint or may make no difference to your life. Be careful what you ask in your wish," warned Papa. Before Victor could reply, Bosco and Sam arrived and the conversation changed to greetings. Mama and Titti were out too and finally, Timson appeared. All sat on the veranda watching the sunrise. The silence at the breakfast table meant no one knew how to start a conversation that involved the name Sofia.
They ate breakfast in silence thinking. It is possible that the sunrise could be a subject of talk but, the name Sofia dominated the thinking. Sam was oblivion to the situation. She kept on talking about the beauty of sunrise without getting any comment from any other person including Bosco. Sam did not detect hostility but thought something was about to take place. Papa, Mama, Titti and Victor including Timson appear to be in deep thoughts. Sam found the silence frightening and she needed to know.
The silent breakfast was over and everyone was waiting to see who would break that silence. They were very alert when Mama said to Timson,
"We are sorry that we upset you. In our family, we have incidents that make us cry as we remember. Victor remembers her mother he lost aged three years, he still cries. Papa goes to his bed, and at times he cries, when he remembers the son we lost. We, however, mention their names on days we choose to honour them."
Timson looked at Mama, then turned his head to glance at everyone and finally said,
"That is your way and not mine. I do visit her at the grave and talk privately to her, hoping she will pardon me before I die. That

is not the same as shouting her name to strangers that gave food and shelter."
Mama replied and said,
"If going to the grave is good for you, I will ask Bosco and Victor to accompany you."
"Why are you tormenting me? Do you want me to go back to the grave and talk to her, accompanied by two strangers?" asked Timson.
"It is Victor and Bosco, your friends, who will accompany you," said Mama.
Timson stood up, and spoke loudly in a temper that alerted everyone as he said,
"You have so far acted nicely to me. Now you are turning out to be exceedingly rude. I will not visit Sofia's grave with anyone. It is not a display ground. Now I will go away and thanks for your kindness but your rudeness has ruined my visit."
On hearing the loud voice in a temper and the words, grave and Sofia, Bosco moved closer to Timson and looked at Mama rather sternly and in a loud voice asked,
"What happened to my mother's grave that Timson does not want us to visit? What has that anything to do with Timson?"
Timson stood still and silent possibly trying to understand what he just heard. Victor stood and sternly looked at Timson and said,
"You may wonder why we are concerned. Sofia, a namesake to your sister, was my grandmother and Bosco's mother."
Bosco was abrupt when he said,
"What is going on nephew? You do not have to explain to Timson about our family."
"Calm down uncle and please listen. Sofia is the name of Timson's sister buried in the village north of Moshiana. My grandmother, your mother is Sofia. She died and her grave is in the same district. It appears a coincidence worth checking. Could it be that Sofia, my grandmother, your mother who died in the

village north of Moshiana, is the same person as Timson's sister?"

Victor saw Timson seated holding his head with his hands while his elbows rested on his thighs listening to the revelations. After a pause Victor then said,

"Timson, would you not want to know whether we are talking of the same Sofia or not?"

The thought was like a spider crawling on Bosco's spine. Frightening, disgusting, tingling or all three sensations together were how Bosco was feeling. The thought of Timson, the vagrant, befriended over a pint of beer, likely to be his relative, an uncle, his mother's brother, was a situation that Bosco could not imagine or expect.

Timson's mouth remained open until Papa closed it saying, "It is better to know for sure." Timson's whole body was shaking. Titti gave him a glass of water. After a brief moment, Timson coughed to gain a voice and then said,

"I do not understand. Bosco, you call Sofia your mother and Victor calls her grandmother. How is that possible when Victor is not your son?"

Victor said,

"It is a long story. We go to the grave and we can talk details or the grave will reveal all."

"I am not a fool. You are a cruel family intending to prolong my agony. My sister did not have two children; I only heard she had a child. They took the child away promptly from her and no one knew the whereabouts of the child. She must have had her young heartbroken by the trauma she suffered. I feel sad that I let her down," said Timson as he stood up. His legs were wobbly. Papa gave him his walking stick fearing he was going to fall. When he was steady, he said,

"Let us go to see the grave of my sister Sofia and then we can go to see that of Sofia, your mother or grandmother. After that, I will go and never visit a family that takes pleasure in insulting the homeless poor. I will never give you a chance to insult me again. Let us go and goodbye all."

Bosco gave a kiss to Sam as Victor did likewise to Titti before they went on a journey accompanied by Timson.

Would it be a journey of revelations or confusion?

On their journey, Victor whispered to Bosco saying,

"I am very worried that there may be two graves for different women called Sofia."

Bosco touched Victor with his sweaty hands and said,

"In that case, all we know may be wrong. To start new research to find our Sofia would be tormenting."

Victor whispered to Bosco, who could hear Victor's heart pounding as he said,

"You are my uncle, there is no doubt but we could have written the names of her children on the wrong gravestone."

Bosco said,

"I know that nephew but I will be terrified if it is the grave we know."

Victor said,

"If it is the same grave, then Timson is likely to be related to us. In that case, I can accept that than to endure more confusion of research by Bevan."

43

A revelation either may be devastating or may bring great pleasure. You could take pity on Victor and Bosco who were very nervous. Timson however, was confused and angry. A long list of possible scenarios occupied their minds. There is no pleasure in taking a trip to visit a gravesite as it could stir up past sad memories.

The trip they were taking was not to an ordinary grave but to one that could have a devastating or mollifying impact on that family. Timson, however, was suffering immensely due to the guilt of failure in his duty of care for his sister. He could have saved his sister's life if he did not give up struggling and instead turned into a beggar and a vagrant. In addition to the guilt, Timson was in the company of two people claiming family relations with Sofia. It would be traumatic as Timson wondered whether it was the same Sofia or not. If it were the same Sofia, Timson thought, Bosco and Victor would blame him more for his neglect. The thought made him feel and look frailer with guilt and unable to talk.

Timson, Victor and Bosco travelled in silence. Timson was leading the way. The closer they approached, the more they became anxious.

Timson's hands were shaky as he arrived at the village. He walked slowly in the direction of the graveyard. Bosco and Victor held each other's hands tightly. The suspense was turning the uncle and nephew into a trance with fear. Timson walked slowly toward the gravestone. The direction was the same as the gravestone of Charles and Charlene's mother. Bosco and Victor followed trembling as Timson approached the grave of Sofia and cried aloud,

"Who the hell added son Charles and daughter Charlene? I did not write that and I did not know such persons." Timson was angry and shaking. Victor's quick reaction prevented Timson from injury as he was in danger of falling on the gravestone.

Bosco moved quickly to help and as the three men stood to look at the gravestone, they realised that Bosco was standing by his uncle and Victor was between his great uncle and uncle. Their wobbly legs forced the three to sit by the graveside, silent with their minds trying to understand what had just become a reality. After a while, Bosco was able to speak and he said,

"I am Charles and Charlene, my twin sister was Victor's mother. You are my uncle and I am Victor's uncle.

Victor was recovering from his shock and said,

"So you are my uncle's uncle."

Timson managed a smile to know he was with family. He could neither speak nor walk. They picked him up to the transport to return home. Bosco and Victor heard Timson in a very weak voice asking,

"Where is Charlene, my niece?" Victor whispered to Timson and said,

"My mother Charlene died in a car accident with my father when I was just three years old." My grandfather, Papa will tell you more.

At a distance, Papa saw Victor and Bosco holding Timson, with their arms shoulder to shoulder. He called Mama and Titti and said,

"Something is not right. Timson may be ill or he may have collapsed. We pushed him too much to reveal his past."

Mama said,

"Papa, look at their smiles, there is nothing wrong. I think there is a surprise, most likely a revelation of good news that accompanies those smiles. Timson had said goodbye but he is here again."

At the veranda, Bosco said,

A Dot on the Timeline

"Papa, Mama, this was my friend and is my uncle and Victor's great uncle and Alena and Sofia's great-great uncle."
"Ok, we understand. You found that there was only one Sofia," said Mama. "Yes Mama, it was the same gravestone where we added her children's names," said Bosco.
Mama touched her cheeks and said,

> *"Sofia must have loved her children immensely.*
> *Many years after her death,*
> *She managed to bring her lost family together again.*
> *Very few beings can achieve that."*

The family sat for an evening meal. No one spoke. No one was sad but everyone was thinking.
To Victor and Titti, Bosco and Sam it was a triumph, they discovered more of the lost family. Papa and Mama were happy that they saved a family member from vagrancy. Timson got an unexpected reward of finding his lost lineage of a family by his act of repaying a good deed. The increasing size of the family was overwhelming. Victor and Bosco who sat quietly on the veranda could not imagine Timson's thoughts. Papa and Mama were equally stunned as they looked at Timson's grey hair, which had shown the grace of growing old. Victor saw a face of the past he could not have imagined. He added the missing past to his short family tree. Victor smiled as he looked at Timson and realised it was his uncle's uncle, a great uncle, with features that Mama and Titti noted and suspected of a possible family link. Filled with emotions, Victor was proud to embrace his past relationship, faced Timson and said,
"You have made my world complete. I have a past and now I can imagine a future. You will have a lot to teach us."
Timson replied saying,
"Victor, I will give you as much information about the past as I can but I have no idea of the future. Finding and meeting my lost

family is possibly a pardon from my sister and I must care for you to repay my past failure."

The voices of everyone appeared released after a cup of tea. The conversation started with many questions and concluded with several old folk stories from as far back as four generations. The excitement of the old stories went on for a long time every evening after dinner. Timson dominated the conversations when he narrated his growing-up days. Victor was pleased to hear the name Donald as the father of his grandma Sofia. He had no images, as they were too poor to take pictures. In desperation to lack of pictures, Victor asked Timson,

"Have you got anything that belonged to your sister that we can touch?" Timson touched his chin for a good while and then moved close to Victor and said almost whispering,

"I am very sorry my great nephew. I had something, an engraved ring; it was among her belongings. I wore it for some years in her memory. One day I met a girl, a very young one, who was nice to me, I gave it to her, and she gave me a necklace as a token of friendship. I never met that girl again to this day and I regret giving it away. I have nothing to show except the necklace I received in exchange."

Victor took pity on his old uncle and said,

"You should not feel ashamed; many a man makes such mistakes."

Bosco took an interest in hearing about a girl and said,

"How long did you live with the girl?"

Timson replied,

"A month or two passed and then I drifted away."

Victor asked,

"Did you leave or were you forced to leave?"

Timson said,

"One day I left home and never returned. Time passed and when I thought of looking for her, no one appeared to know her and my life as a drifter persisted to this day."

A Dot on the Timeline

Victor whispered to Bosco,
"There could be a woman looking for Timson. I wish she found him to make him happy.
Bosco whispered back and said,
"Based on what happened, I will not rule that out."

44

Papa and Mama as they rested after dinner recollected the incidents that led to an increase in their family from three to nine. The increase, gave a sense of achievement.
Mama said,
"I thought, for a long time that our family was just you, me and Victor forever."
"Don't forget Titti giving us Alena and Sofia," said Papa.
"Can you imagine a quarrel between two people led to the discovery of Bosco?
"I was about to get a heart attack when Victor thought of Bosco as a possible uncle," said Papa with a smile. Mama after a long pause said,
"Sofia must have great love. It is only love that can bind a family strong enough to bring her family to one dining table after fifty years."
Papa turned to face Timson and said,
"Timson, you were the missing link. You have cleared all the past unknowns and given a glimpse of the life of the past."
Talking of wishes, Mama witnessed her aspirations becoming true but not as imagined. Papa had a man of his age to talk about the glorious old days. Victor, Bosco and Titti heard tales of wisdom that at times appeared unbelievable but gave them a past that they then could not imagine. The old stories, the revelations of characters of past family members, the jokes, the laughing and the questions continued night after night. The younger ones, that included Bosco, Victor and their families, had their past as clear as spring water and their excitement as high as cirrus clouds. The good mood of the family was bound to come down sometime. In life, difficulties are bound to occur, thought Victor.

A Dot on the Timeline

Amidst the laughter, there was a cry. The laughing stopped. There was a sudden sense of fear. A cry warns that there is a problem. It was a cry of pain by Sam. The men had no clue what was happening. Mama and Titti quickly took her inside. Mama remembered those pains vaguely but the urgency of acting in such circumstances was very clear to her. Titti remembered the sequence of the series of contractions that led to delivery. They had to call a knowledgeable woman. Sam was in labour. She needed calming and she was safe. Papa, Timson, Bosco, Victor and the children Alena and Sofia sat outside. They had to answer constant questions from Alena and Sofia. Questions like "What is happening to Sam?" were easy to answer but the question that followed, "Why?" was more difficult to give an honest reply to avoid a myriad of more *'why'* types of questions.

Waiting is like a sleepless night when the clock's tick-and-tock noise appears never to end.

At midnight, a call to Bosco broke the silence. Bosco was nervous. He stood up and ran towards the room as fast as a schoolchild summoned to the head teacher's office. He opened the door. There was a cry followed by laughter. A son was born. Bosco did not stay in that room for long. He walked back to meet Papa, Mama, Timson and Victor. He did not have to say anything but stood watching Victor opening a bottle of wine until Papa asked him to sit. After a glass of wine, Victor congratulated Bosco for becoming a father and said,

"We thank you, uncle, for making this family bigger and happier. I thank you for giving me a cousin."

Timson stood up without needing his walking stick. He walked slowly to Bosco, gave him a heart-warming embrace as he said, "Now you have a son, I am pleased that I will not die before a person calls me great uncle again. I can see you are filled with pride when Victor calls you, uncle." Victor and Bosco helped Timson to walk to his chair. After serving a glass of wine to everyone, Victor smiled looking at Timson, Papa, Mama and Bosco and then said,

"Sofia was my grandmother and Donald was Bosco's grandfather. Would it be suitable to call the baby Donald?"

Bosco did not reply but went straight inside and then returned smiling and jumping like a frolicking fox and shouted,

"The baby's name is Donald." The celebrations of the birth of Donald Justin went on for a week. The rejoicing was awesome. Donald was born into a loving family protected by a father, a mother, an uncle, a great-uncle and cousins. Papa and Mama with Sam and Titti's parents were the grandparents Donald would grow to know. The family was wondering what else to do before the planting season. Ten family members occupied Papa's dining table. You will think of a family as a bubble and if it is, it will link to other bubbles. Victor thought and assumed the chance of finding more family members was a slim possibility but not impossible.

Bosco was learning how to be a dad. He did not do well in classroom education. This time, however, he has to succeed, as no one will tolerate failure. Parenthood is a school that goes on for life. It will be a miserable existence to live with that kind of failure. That was a worry in Bosco's head, whenever he heard the word, dad.

Victor wondered how Bosco who started late, as a parent would cope with the energy demanded by a boisterous youngster like Donald.

45

One evening as the family relaxed, the younger members sat quietly expecting a hilarious story of the past from one of the seniors. It was just after sunset and a young woman was approaching the family house. At earshot, she asked aloud,
"My name is Harriet. I am looking for a friend called Sam, am I at her residence?"
"Come in please," in unison Sam and Bosco said. Harriet smiled and walked towards the veranda faster and more confidently. After the greetings and some pleasantries and introductions, Titti added the eleventh chair for Harriet in that evening meal. It was temporary they thought. The table was big and Papa and Mama were proud. After dinner, the whole family went to relax on the veranda that had relaxing soft chairs and Papa's rocking chair. Timson whispered to Papa and said,
"This is the daughter of the man killed in the bank raid, twelve years ago." Papa realised that Timson and Bosco had close encounter in the past at the bank and that Timson protected him. Papa was quiet. After a moment of assessing the situation, Papa realised that Timson had protected Bosco at all times even when Papa asked how they met. Papa looked at Timson and admired his devotion to his family and the love he unwittingly showed to Bosco before he knew he was his nephew. Harriet thought that Timson looked like the man who asked them to go outside of the pub to meet Bosco. She shared that thought with Sam. Sam remembered the pub incident and the many questions she intended but postponed to ask Bosco. Bosco moved closer to Harriet and said,
"I hope you remember Timson. He is my uncle. It is a long story to be told another day."
Harriet replied,

"I used to see him around but never thought he knew you."
"We did not know one another until the recent incident that I will tell you later but first there is something I want to ask,
"Sam has not told me much about you since we met. Can you tell us about your family?" Bosco intended to find out if the Harriett suspected anything detrimental to him since they hurriedly left Moshiana.
Sitting around such a big family made Harriet nervous and she coughed to clear her throat and then took a sip of water before she spoke.

"My father was a manager of a coal mining company. He was well paid and we live comfortably with my mum. Ten years ago, a gunman killed him and life at home was hard. My mom could not work. She was a broken woman, unable to look after herself. She died after five years of heartbreak and I was left struggling alone until I got a job working with Sam."
"Sorry, to hear that but your father was probably the one who gave me my first job in Moshiana. When I saw his picture, he resembled my adoptive father. Did he have a brother?" asked Bosco.
"Yes, but they never spoke because my father blamed him for taking all the inheritance money from their father. My father said that there was a curse because they did not have a child. I heard my mother telling my father that his brother had to adopt a child to leave the wealth to him. However, I never met the boy described as fat, conceited and looking a bit stupid. I did not ask to know his name."
Bosco's heart raced. Some facts matched his childhood; other facts could reveal his near criminal past. He remained silent. Papa, Mama, Timson and Bosco knew of the near criminal past but Victor, Harriet, Sam and Titti did not know the whole story of Bosco past. Harriet worried at the silence that followed asked, "Did I say something rude?"

A Dot on the Timeline

Victor went close to Harriet, spoke in a low tone of voice, and said,
"This family has witnessed many surprises. You are probably going to give or hear a surprise. Nothing will surprise this lot."
Harriet said,
"I have no surprises; it is the truth of what I know of my family."
Victor had to think fast and then asked,
"What is your family name and do you know your uncle's name?"
Harriet looked at the face of everyone and found that all were looking at her. She sensed a kind of hostility and stood up, went close to her friend Sam and she said,
"I think I had better go. I think your family does not like me."
Sam stopped breast-feeding and said,
"Please listen to Victor, he meant well in his questions."
Harriet turned to Victor and said,
"I came here because Sam invited me but your questions are too much and too soon for friends to want to know. However, I will tell you and then I will go."
Victor said,
"I mean no harm; it is curiosity from the picture I saw."
Harriett picked up her handbag, smoothed down her dress ready to go and said,
"My uncle was Boniface Justin and I am Harriet Bernard Justin, Goodbye."
She turned and looked back when she heard several sounds of breaking cutlery. Cups and saucers slipped from many hands, as there was a realization of a family connection. Bosco called out loudly,
"Please stop and listen. I am the fat, stupid and conceited adopted child. I am Charles Bosco Boniface Justin." Harriet stopped; Victor rushed to ask her to return by holding her hand. She sat to listen and she heard enough to make her cry with pleasure when passes an exam unexpectedly. Sam was surprised to find that Harriet was her husband's cousin. Bosco was

emotional. Victor was wondering how the small family bubbles were merging rapidly to form their large family bubble. Harriet was too overwhelmed with thoughts that kept her crying until Sam moved close to offer solace. She stayed the night.

In the morning, the whole family sat to enjoy breakfast. They talked freely and joyfully knowing Bosco had a cousin.
Timson approached Harriet and said,
"I remember you at the pub. At that moment, I did not know that your friends were my relatives. I came for a visit as they invited me. They acted in a good turn to what I did at the pub. When I arrived and told them about me as you did, they asked me many questions as they did to you. I was angry and I wanted to leave when a name united us, as it has done to you." Harriet relaxed and realised that families have something that tends to unite them and thought that she was probably part of that process. She was happy and her planned short visit became a long stay.
She never left until Donald started walking. After that, other reasons beyond her control made her stay longer abandoning Moshiana. It was uncertain whether Harriet would leave a loving family of eleven members to a lonely flat in the dingy crime-ridden town of Moshiana. She opted to stay with the newfound family. She had a friend, an adopted cousin she discovered and the offer of a new job at the township bank. There were more incentives but she did not tell.
The decision to stay was proof to Victor that his family was a happy place.
Victor was proud as he reflected on how Bosco, Timson and Harriet joined the family. In that thought, he recalled the words his grandfather had told him in his teenage years.

A Dot on the Timeline

In life, you may be poor or rich
Do not sneer or be harsh to the less gifted
Some actions can hurt, choose good acts
Good deeds may turn you from poor to rich.
You will never know until you try

Papa looked at his family seated. He counted. He was satisfied with the quantity of eleven that was likely to grow. In a contented and an elated mood he said,
"We are now strong as any family with a bright future. Scattered and alone, in various places, we were unaware of any living relatives. We should aim for a future of unity and glee."

The Future Assured

46

You could laugh at Victor or call him unwise. That would be your opinion but he was proud that he failed to get a high-flyer job in the city. You would change your mind when you hear the reasons that make him proud despite the failure. He chose to be with his family to continue farming. That wish, with the added benefit of discovering an uncle, made that choice a very wise deed.

He remembered his quarrels with Bosco when the relationship had stretched to dangerous limits. He smiled knowing the challenging times he had experienced were over and was proud because he had so far sailed through successfully.

The discovery of an uncle, a great uncle and an uncle's cousin gave Victor buoyancy to sail over future problems. He was almost certain that such discoveries were unlikely to happen again.

Sitting on the veranda, looking at Papa and Mama, swinging slowly on their rocking chairs and enjoying the early morning sunshine, Victor thought about the constant arguments he had with his teenage daughters. He wondered how Donald, Bosco's son, would turn in a year. No one could predict the change in character that takes place in the teenage years. He remembered what Bosco told him about backchat from his child. The thought that he could take it as seriously as a terrorist act worried Victor. He was grinding his teeth. He was angry at his failure to guide his teenage girls. Papa was not asleep. He saw the grief of his grandson and said,

"You worry too much."

Victor looked at Papa and said,

"I cannot talk calmly to Alena and Sofia. They are abrupt and at times rude to their mother and me. Did I behave rudely at that age Papa?"

"You were not an Angel," said Mama.
"What did I do?" asked Victor.
"You were playing with your grandad's eyeglasses and you set the hay into flames."
Victor scratched his hair and said,
"I cannot remember that incident."
"Papa called you to look for his eye glasses. You must have panicked and left his glasses by the bale of hay. We saw smoke and quickly extinguished the flames. You returned to see the bales on flames and said nothing. It was when we found the eyeglass," said Mama.
"Now you see the children may be afraid of something similar as you were afraid of telling," said Papa.
"Give them confidence, they may surprise you," said Papa.
"They are teens, they see life, not like us," said Victor.
"Try to listen to their reasoning. You may remember your days at that age," said Mama. Papa grunted and said,
"At times, you may need a firm stand. You cannot aspire to appease all the time. The world is rough and unforgiving. It does not show mercy when basic rules of existence are disregarded."
"What basic rules do you mean?" asked Victor.
"An example is that of a good turn deserves another," said Papa.
"Do you think they will understand that?" asked Victor.
"Do not worry you have a perfect example to use to make it clear," said Papa, as Victor nodded in agreement.

Timson was sitting on one of the many steps that led to the veranda. He turned to look at Victor, smiled and said,
"Welcome to the world of adulthood. Children like to get praise. You give them compliments for anything good they do and you note their reaction."
"I have nothing of praise to say as all their answers are rude comments," replied Victor.
"Do not take every act as rude," said Mama.

"How shall I achieve that?" Victor asked, raising his hands as a man defeated in a duel.
Papa saw that Victor was about to accept failure. He coughed to attract attention to speak but Bosco walked in and greeted Papa, Mama and Timson and then said,
"May I speak to Victor please Papa?"
Turning to face Victor, Bosco said,
"Nephew we have to talk,"
"What's up, uncle?"
"We have missed three dinner meetings, Papa and mama might find it unacceptable. They deserve an explanation."
"Papa had noticed that absence and said we wait to find out the problem," said Victor.
"Yes, it is a problem. I have no solution yet as it involves both of us," said Bosco.
"What is it?" asked Victor.
"Sam thought it was best to keep away. She thinks that Titti did not like her because she said she stunk of sheep muck," said Bosco.
"I do not think Titti would talk like that. She is a farmer. She knows smells and muck," said Victor.
"I thought that too but we have to find the source," said Bosco.
"I will talk to Titti," said Victor.
Papa coughed again and spoke saying,
"I am glad to see you trying to prevent quarrelling in your families. It shows you are acquiring wisdom."
In unison Victor and Bosco replied,
"We will not let feuds split the family."
They noted Papa nodding in agreement but then they heard,
"Boys, you know I am a super snooper," said Timson.
Bosco and Victor turned to face him.
Bosco asked,
"I know you are the best in spying but what do you know that we do not know, uncle?"

"At the ages of twelve, fourteen and sixteen, children start knowing things. They repeat profanities they hear without fully understanding. They fight to show strength or to attain dominance."

"We know that, what is new?" Bosco asked in a tone of voice that revealed some tension.

"Cool your temper, my nephew, I will tell you," said Timson smiling.

"Please great uncle, tell us. We are on tenterhooks," said Victor.

"The problem is not your wives; it is the children's talk, chit-chat, games and play that went a bit astray."

"What kind of game?" Victor asked rather abruptly.

"Children play and at times they argue. It is in one of their arguments that the stinking issues made them raise their voices," said Timson.

"Who started it?" asked Bosco.

"Do you expect me to tell?"

"Sorry uncle, we did not mean that," said Bosco.

Timson said,

"You know I have lived without a fixed aboard for a long time. Many people think you are stupid, ignorant or at worse with a mental illness. Most vagrants have none of those conditions and they could be clearer-minded and wittier than most proud smartly dressed people could. They solve living problems in a way others do not understand."

"I believe what you are saying, uncle. I have proof that your memory is sharper than mine. You remember details of incidents and faces, better than I do. I am sure of that," said Bosco.

"At times I think deeply and listen to others in the same predicament of life. The wisdom I hear makes me wish to remain in that community of vagrants," said Timson.

"What kinds of stories do they talk about?" asked Victor. Timson was proud that Victor and Bosco were keen to hear his story, which would distract them from the immediate problem.

Bosco and Victor were prepared to listen to a story about the life of vagrants.
Could the story help to solve the initial problem?

47

Papa and Mama were keen to hear Timson's story too, as all sat in silence.

"I will recall a story by one vagrant called Beer-gut. He got that name because he could swallow a whole pint of beer in few gulps before a pause to breathe," said Timson.

"Beer-gut did not start as a vagrant but he was a combatant. He had been in many wars and learned just one thing, obeying orders. There were no ifs or buts associated with that command from your superiors. Disobedience was risky; it could result in death to you or your team. A change of orders may be unsafe and result in death or disciplinary."

Beer-gut would say,

"Those who give orders are in secure places and far away from the conflict zones. After a long conflict, many would forget the purpose of the fight. You cannot talk to your opponents and ask why you are killing one another but the generals who give orders could establish communication. They do not make that attempt despite many opportunities because they are not in danger of dying. In the event they were in the frontline, their attitude would be to put more effort in seeking peace than to fight."

Beer-gut with his mouth twisted in disgust said,

"The life of another person is cheaper than one's own." Beer-gut went on to say that, if he were a general, he would go in the front with his fighters and ask the other opposing general why they are fighting. He would probably answer,

"I don't know."

"Shall we pause and talk?" Probably such a gesture could end the conflict and find lasting solutions.

Impressed by his wisdom of the story, I dared to ask why they called him Beer-gut.

Twisting his mouth as a person who tasted something bitter, Beer-gut said,

"My friend, in the conflict zone we expose ourselves to danger and try the survival mode by killing anything that moves in the sights of your periscope. Things you see are not good for the human mind. When you survive and return, what you saw, repeats in your head giving stress to such an extent that you cringe. Drinking beer momentarily reduced the stress but I drank so much that I could not get drunk anymore."

"Why did you choose to be a drifter?" I asked Beer-gut again.

"I wanted the freedom to do what I want. Free from taking or receiving orders," said Beer-gut.

Timson finished his vagrant chat story. He looked at Victor and Bosco, He scratched his bald patch and then he said,

"You have behaved like the generals. Your children see that attitude. You have to show that you share their struggling with them and show empathy in their pain. The children are observing so many changes taking place in their bodies and surroundings that are frightening as the mind opens up to many life realities," said Timson.

"How can we do that?" asked Victor.

"Do you remember your worries as a teenager?" asked Timson.

"Not much," said Bosco.

I can remember how worried I was when I woke up in the morning and spoke with a low tone of voice like Papa," said Victor.

"What did you think it was?" asked Timson.

"I did not understand until Mama said that it was a sign I was growing up to be a big man like Papa," replied Victor.

"Your children, at that teen's age, have no idea that you ever went to school. Therefore, they think you do not understand their problems. When they talk to you, they do not expect you to understand their plight. Your answers may be the cause of the rebellion," said Timson.

"We try to offer solutions," said Bosco.
"Your solutions are not as good as their solutions," said Timson.
"How can we know their solutions?" asked Victor.
"Do you understand what they see in you?" asked Timson.
"May be as a person who gives orders," said Bosco.
"How do we ensure their thoughts and ours are mutual?" asked Victor.
"You talk, by asking their views; the generals did not," said Timson.
"We talk to each child," said Victor.
"That tactic of authoritative encounter will not work," said Timson.
"Do you expect us to reward them? Victor asked irritably.
"Bosco should do the talking with all the children while walking in the sheep farm, not sitting across the table like in an office. The office setting will remind them of their head teacher's office and that is when they have done something undesirable. It will amaze you when the children are relaxed especially if you let them talk. Teens find rebellious behaviour a thrill when not listened to," said Timson.
Papa raised his head and said,
"Ask them to be the boss for a while and see their actions."
"We will do our best and report to you Papa," said Victor.
"Good luck," said Mama.
Bosco scratched his head and said,
"We should start by introducing the children to farm chores."
Victor nodded and said,
"We should work with them, learn a few of their thoughts, and then we make enquiries. We do not ask questions until at the end of their holidays,"
"Donald will go to sheer the sheep with me," said Bosco.
"Alena will plant grapes with her mother and Sofia will plant potatoes with me," said Victor.

"In the evening, the children are paid but their mothers to keep their wages," suggested Bosco.
When Sam received Donald's wages, she said to Bosco, "Your son is asking many questions about sheep and pigs."
It is said that time heals many ills. The rift between Sam and Titti over a comment that one stunk like sheep muck, appeared forgotten as communal evening meals resumed. At those evening meals, there was happiness when the children joined in the talk about the farming tasks they had done. No one complained or claimed boredom when Bosco gave a talk on the value of farm work. Bosco and Victor claimed victory. They had solved the problem without resorting to questions and answers, threats or punishments.
The school holidays had ended. The children returned to school aware of the rewards of work and responsibilities. At evening meals, the communication between parents and their children increased. Bosco and Victor in privacy with Papa, Mama and Timson gloated that they had succeeded in guiding their children and claimed that they would never fail again. They got a nod of approval from the three seniors.

48

At Papa's table sat eleven people, old and young. The old ones required care while the young ones prepared to get into grips with the needs their care demanded.

All, the young and the old, said goodbye to the old year and celebrated the arrival of the New Year. The young were aware of the worsening health conditions of Papa, Mama and Timson and assured them of safety and comfort. Bosco, Victor, Titti and Sam with Harriet discussed and agreed on care plans. The seasonal tasks on the farm went on as usual. In times of worrying episodes about the older relatives, Victor said to Bosco,

"I will not survive without the wisdom of Papa and Mama. They are about to bow out."

Bosco said,

"We have to care for them, to the best of our ability. We have to learn to stand on our own at the same time."

Titti heard the exchange and said,

"Do not worry too much. Whatever happens is a dot on the timeline, written for them and that is meant to happen."

"Timson will fill the gap," said Bosco.

Victor replied and said,

"He is alert mentally, but cannot do what Papa and Mama could do."

"You are correct," replied Bosco.

"A year always starts and finishes in the same way. However, what happens during the year can be rewarding or traumatic," said Victor.

"What do you mean, nephew?" asked Bosco.

"I was just thinking aloud. Pardon my stretched imagination of the future," said Victor.

The sun rose, the birds chirped, and Victor with Bosco met for farm inspection as usual. Titti however was feeling unwell and stayed in bed longer than usual. Timson woke up to check on Papa to tell him a story he remembered in the night. Papa responded to the morning greetings and said to Timson,
"I feel well but a bit tired. I will stay in bed. Call all the family to tell them the story. I will open the door and listen to it." Timson suspected that Papa was unwell. He called all and as they arrived, he said,
"I think Papa is unwell because he did not come to the chair at the veranda as usual." Victor, Bosco, Titti, Sam and Harriet sat around Papa's bed. Mama saw their arrival and said,
"Papa is well he had a good sleep," Papa replied and said,
"I can speak for myself. I told Timson that I am tired but not ill."
Timson said,
"In my days as a vagrant, when a person says they are tired, it was not taken lightly. I am sorry I frightened the family. I will tell my story another day." Titti provided a cup of tea to all. Mama was drinking her tea but Papa said he wanted to rest a bit more. He dozed off as usual. Mama was drinking her tea while the other family members were exchanging their plans for that day.
"Victor," said Mama, "I think your grandfather's tea will get cold." Victor stood and touched Papa. Normally he would wake up but that time there was no response. Mama thought the worst, she moved closer and in a voice that said it all made everyone stand and become silent. It took a few seconds that appeared like hours before tears, sobbing and eventually loud cries emerged from the room. Timson was strong and able to move everyone to the veranda and start preparations that had to follow. It was a change that everyone knew it could happen but never dared to admit. The thought alone was traumatic.
The passing of Papa marked a traumatic moment in the timeline of the family. To Victor, Papa was the only mentor he knew. To Bosco, Papa was the man who guided him to happiness. The

void caused by the death of Papa did not only affect the close family but there was agitation and fear in the village. Apart from fear in the village, Victor had his grief.

> *Papa was probably ready*
> *I was not prepared, for the shock*
> *To live without his presence*
> *Hope time will aid the healing*

Timson was not new to trauma. His friend was Papa and the only person he trusted and respected. The void made him vulnerable to worrying thoughts like the loss of his sister and the guilt returned. Timson frequently walked on the farm alone at times talking to unseen companions. Bosco noted his uncle's dangerous state of mind. He told Victor his worry saying,
"My uncle is losing his mind and I see his will to live, appear to weaken. He is unwilling to chat as before."
Victor said,
"I am aware and worried. We should care more and increase monitoring his whereabouts." The mornings were becoming tense moments for the family. The worry became real one morning when Bosco went to check on his uncle, Timson. He had eaten well the evening before, complained of nothing and went to sleep at the normal times that became the sleep for eternity. The trauma of the passing affected all but Bosco had lost a mark of his past he came to know later in life. With his family behind him, he underwent healing but it was a slow process.

49

Mama was too frail to take charge of the family matters. There was fear of the direction Bosco would take the family in the absence of Papa and his uncle Timson. The loss of Papa appeared to have taken all the strength from Mama reducing her to depend totally on the care of the younger family members. She had to give up as the decision-maker of the family and asked Bosco for his opinion. She wanted a smooth transition of family leadership and said to Bosco,
"You are the eldest and the two farms have performed well as two linked businesses and should have one leader to ensure future prosperity. What is your opinion?"
Bosco was in a dilemma. He could neither accept nor refuse Mama's request. He thought that acceptance would put the farm in disrepute because of his past deeds.
He said to Mama,
"In the village, my reputation is not good. The neighbours may not have forgotten my past or may not be ready to forgive me. I will support if Victor becomes the leader and I will give him my full support." Mama was proud of Bosco's opinion and said,
"You have put the success of the farm before your ego. It is a credit to you. You have acted truly in the protection of the family. Talk to Victor about the leadership."
If you think the transition of authority in families is simple, then you need to think again. A process starts with fear and distrust. That situation can worsen through greed or outside influences and that will test the family's strength of unity. The changes, notably the passing of Papa, brought uncertainties to the neighbours and worries to the family.
An early morning knock on the door frightened but woke up Victor. He opened his window and asked,

A Dot on the Timeline

"Who is it?"
"It is Goodwood; I want to talk with you."
"Can it wait?"
"No it is urgent," replied Goodwood.
Victor threw his bed covering to the side, put on his robe and went to meet Goodwood, who said to him,
"Sorry for the disturbance but there is great unease in the valley. The neighbours assumed Bosco would take Papa's position in the family. There was a widespread fear that he could revert to his old unsocial behaviour. Many farmers need to know that the assurance of a peaceful existence your Papa promised would be honoured."
Victor had not forgotten the old days when Bosco used to terrorize the farming community but he was almost certain that he could never change into doing those old unsocial practices. He had a nephew, a wife, a cousin and a child and if he had not become wise, heaven help us, Victor thought.
He patted Goodwood's shoulder and said,
"There will be no repeat of those old bad days of Bosco's unsocial behaviour. He is my uncle, he is a good man, I promise."
"What assurance can you give?" asked Goodwood.
"The pact given by Papa is solid and Bosco cannot change it," said Victor.
Goodwood left but he was still worried. Could the young man control Bosco? He thought of possible challenging issues again shortly but left trusting Victor's promise.
You cannot ignore the news you read or hear. Victor could not relax until he was sure whether Goodwood was right or wrong. He returned to his bed but could not settle. He was worried. Bosco could attempt to take over the leadership of the family and may try to order him around, a situation he dreaded. He had to know Bosco's intentions. A feud with his uncle was a situation Victor did not want. He dressed and went out to the farm in the hope that Bosco would introduce the leadership talk. Victor saw

Bosco and said, "Have you ever had a frightening dream that you had to tell someone in case it could happen?"

"No nephew, I never dream but most times I was angry and I did not know what made me irate."

"I think it is something like that," said Victor. Bosco looked at Victor wondering whether the shock of losing Papa and Timson was affecting his nephew. He patted his shoulder and said,

"Some dreams could be a communication with the departed loved ones trying to guide you to safety."

"You may be saying something believable but I do not think that happens because dreams are totally in our minds not like a lecture when we listen to someone speaking," said Victor.

"How can a dream make us jump or shout or cry," asked Bosco.

"I do not know how but I do not think there is any other being involved to cause the fear aroused in dreams," said Victor.

"It is all our imaginings, my nephew. Let us sit somewhere comfortable and tell me all that you have pictured in your mind," Victor moved to a mound formed by a huge root that pushed upwards. Sitting on the root trunk, Bosco and Victor could see the extent of the farm.

Victor said,

"When I was a little boy, I remember Papa used to sit here for hours. Look at the spans of the farm and probably make plans for the year."

50

Victor sat on the tree root as he remembered his departed grandfather. He recalled one thing the grandad said to him, "Remembering the good, the witty and the kind things the departed did, could reduce the sadness, fear and insecurity usually associated with their passing." He was in deep thought and did not hear when Bosco spoke. Bosco repeated after a touch on Victor's shoulder saying,
"It is a good view that Papa must have enjoyed," said Bosco looking at Victor probably expecting the dream story.
Victor said,
"I do not think it was a dream but Papa featured greatly in my mind. Last night when I went to bed, I remembered he said that my father worked very hard on the farm. When my father met his premature death, Papa dedicated the farm to him. When I graduated, I planned to leave for a job in the city. Papa used all his powers to convince me to stay."
"Thank goodness you stayed. I would be here alone, behaving unsocially, making trouble for everyone and eventually becoming crackers," said Bosco.
Victor appeared to ignore Bosco's words as he continued saying,
"It was after a while that I understood what my dad liked to do. That made me happy and I relaxed and possibly slept, as I was happy I was doing as my dad wished."
Bosco faced Victor. His beard was long and changing from black to grey. He was getting used to pulling and twisting it, especially when he was under stress and then he said,
"Nephew, Mama asked for my thoughts on leadership in the family. I told her that my previous reputation could harm the farm business. Many people tolerate me as I am a member of Papa's family but they have not pardoned me. If I take leadership,

we may lose friends and business. You have to take Papa's roles and I will give full support."

Victor was silent but happy that the worries expressed by Goodwood could not occur. He did not say it but thought that Bosco was showing maturity and wisdom. It was unlikely that he would go against Papa's bond to the farming community. After a short pause of thinking in silence, Victor said,

"I do not think that is what the dream said but uncle we will work together to make Mama happy and the departed contented."

Bosco expressed his happiness by saying,

"We will succeed to have something of pride to leave for Donald, Alena and Sofia."

That support was a surprise but pleased Victor who simply replied and said,

"We will work together as before without showing who took Papa's place."

To show solidarity, the uncle and nephew started to walk around the farms and discuss their views of the future.

Deep in his mind, Bosco had a feeling of losing power. He decided that he would not allow Victor to make all the plans without his say. He was the uncle he had to guide the young leader. That secret could cause problems.

Bosco looked at Papa's uncultivated land and he made plans but did not disclose them to Victor.

All sat at the table for the evening meal and were happy as usual until Bosco stood up intending to talk. The adults were probably thinking about the choice of the leader for the family. Sam and Harriet would want Bosco to be the leader and Titti would prefer Victor, and that is what would happen if there were a vote.

Titti stared at Bosco who stood up to speak after the evening meal. She thought that Bosco was trying to give orders but she calmed at the end of the short speech that acknowledged Victor taking the leadership of Papa. Victor stood to reply saying,

"I will try to guide this family to prosperity for all. I need help and cooperation." Titti, Harriet and the kids clapped and mocked Victor by calling him, Papa. Victor was contented that the family was happy and united but as a man who had experienced many surprises, he was always alert and worrying over challenging times ahead. Victor thought that Sam appeared unhappy with the arrangement. She did not speak, smile or comment but found an excuse to leave earlier than usual. Bosco was also perturbed and followed Sam home.

Bosco and Sam sat by their kitchen table, and then Sam said, "Giving Victor the authority makes you a weak man unable to take charge giving responsibility to a boy." Bosco was not happy that he gave up leadership roles to Victor so easily but he said to Sam,

"We arranged that no one will know who is in charge and that means I can make decisions and he cannot reject but support. I agreed to acknowledge his leadership to use Papa's respect for the business."

Sam said,

"In that case, we can use the uncultivated land to graze our animals. We hope he will not object." Bosco thought for a moment and without admitting that it was not a good idea to act without asking Victor, he agreed to Sam's proposal.

Any new leader will have some insecurity issues. Victor was aware of glitches in his leadership but he never thought Bosco could cause problems. Bosco was on the verge of behaving as badly as in the old days. He expected Victor to accept Sam's plan. Victor had to remember Papa's routines, his values and his wishes for the family's future. He wanted to be as good as Papa was. He had to be careful in what he said or did to avoid family feuds. He knew that it was very easy to start a dispute but to end it was anything but easy. A single word said in the present time or a word of the past repeated is enough to start hostility.

51

Bosco was in bed thinking about the farming business. He planned to speculate to accumulate. He had an idea that would just do that. He had to share that thought with his nephew. Early in the morning, Bosco went to wake up Victor to tell him of his brilliant idea. After a cup of tea, Bosco said,
"Last year's business was good. I plan to increase my stock of sheep and pigs as I got rumours that there may be food shortages and that means price rises."
"What kind of rumours did you hear?" asked Victor.
Bosco said,
"A government worker inadvertently said that there may be conflict in our region. One country is accusing its neighbour of breaking agreements they made many years ago. He went on to say that if talks fail there is a possibility of armed conflict that may lead to disruptions of many services leading to shortages of commodities and hence price rises."
Victor encouraged his uncle by saying,
"I am proud you are thinking ahead. I will support you because when you succeed I am successful too. What are you planning to do?"
"I am a forward planner, my nephew. Profits will be enormous. I will be richer," said Bosco smiling as a student who passed a test.
"Uncle, when you and I see the profits, we can then start to rub our hands with glee," said Victor.
"Indeed my nephew, indeed," said Bosco rubbing his hands.
They were in a jovial mood as they walked to carry out their usual farm inspection.

Bosco was a doer; however, Victor was a thinker. They differed greatly in the methods to follow to achieve the predicted great profits.

Victor thought of war. He had not experienced any conflict but he had read of the devastating consequences of conflicts. He knew that a few people got riches but many lost their wealth in periods of conflict. To Victor, his uncle could be a winner or a loser. In conflicts, price rises are not pleasing to everyone but other complex factors of supply and demand mean profitability is not always guaranteed. The thought discouraged Victor. He decided to wait and see. Bosco however looked at the higher profits only and assumed it was plain sailing. Bosco interrupted Victor's thoughts by saying,

"I will allocate funds to increase my stock of animals expecting to sell at elevated prices. You better do likewise."

Victor scratched his hair and said,

"In the event, there is an armed conflict then the leaders could behave like the generals in the story of Beer-gut, that uncle Timson told us. Those protected generals did not care about the lives of others. I cannot see how they would promise anyone profitability in business," said Victor.

Sam and Donald arrived from their shopping trip. The two men stopped discussing business. Sam did not talk. She was as if she was in shock. She held the receipt of the goods in front of Victor and Bosco and allowed them to read.

"See now nephew, prices are all up," said Bosco.

Sam eventually spoke saying,

"The cost of living is up. We will not be able to afford most good foods soon."

Bosco said,

"We sell more sheep."

"You can sell the whole farm but you will spend the money in one shopping trip at this rate of price hike," said Sam.

Victor asked,

"Has the war started? What will life be like, if it starts? Wars are never good, uncle."

"Look at the profits my nephew, you worry too much," said Bosco.

"For now uncle, I will do what I have always done. I will change based on the effects the conflict creates."

"If you wait, it may be too late to harvest the profits that will fall like the manna from heaven," said Bosco rubbing his hands.

"I will sell more straw to the livestock farmers who have little land space for fodder," said Victor. A shake of hands sealed a deal to agree to act differently as Bosco and Victor walked to their respective houses. That day things were good as Victor chose not to take risks while Bosco decided to dive in for just one thing, the high profit predicted.

Bosco started buying livestock. He spent a lot of money. Victor made as many bales of hay as he could and stored them in his barns in anticipation of increased demand.

Bosco had overlooked one hurdle. The increased stock needed more land to forage or needed more food. Costs escalated. In desperation, Bosco let the animals roam to Victor's land to forage without consultation. Victor was about to go for his usual walk for farm inspection when he saw Bosco.

"Good morning nephew," he said as he approached.

"I have let the sheep graze on the uncultivated land. I hope you do not mind." It was early in the morning. Victor had not had breakfast. He suddenly felt hungry and started to sweat. Bosco's sheep were eating Victor's stock of hay. He did not reply but asked Bosco to join him for breakfast.

Victor, in a controlled but angry voice, asked,

"What did we agree on decisions taken before action?"

Bosco looked at him smirked and said,

"We make resolutions together."

"Did we discuss letting sheep graze in my land?"

"Nothing is growing there and the sheep need fodder," said Bosco.
"I use the land to grow grass for making hay for sale. You can use it and pay me to match the hay I would sell."
"We are family nephew; we do not charge each other."
"You want to make a profit, I want profits too. We cooperate but not at the expense of the other," said Victor.
"OK, I shall pay you for grazing my sheep, but how much?" asked Bosco.
"An equivalent of twenty bales of hay each day you graze," said Victor. "Extortionate. That is not mate's rates," said Bosco.
"As a leader, family profits must not be compromised," said Victor.
"You may be the leader but I must agree to a decision," said Bosco.
"Since we have not agreed, we are not going to graze on my land until you pay," said Victor. The raising of voices by Bosco and Victor had attracted the attention of Titti who was in bed. She got out and said,
"Are you two arguing before Papa's body has time to settle in the ground? What is the contention?"
Bosco usually left in a huff when defeated, but he did not do that due to Titti's shriek. He looked at Titti and said,
"Your husband is behaving childishly." He walked slowly as a bear that had eaten to its fill to go home. Victor stood remembering the old days of Bosco and saw a feud starting slowly like smouldering amber that eventually becomes a fire.
Titti was enraged and was about to say something that was surely not going to be pleasing. However, Victor's hand warned her to stop. Victor explained to Titti what Bosco was doing. Titti was vexed. Victor's approach of politeness like that of Papa would not do. She walked, dragging her boots onto the ground. It was her expression of anger. She went to see Sam and after pleasantly delivering greetings, she said,

"Are you aware that your sheep are on my farm grazing on the hay that gives us income?"
Maybe Bosco arranged with Victor, but I will ask him."
"No, do not ask him. Bosco would not and could not act like that. Tell your helpers to drive the sheep out quickly before we suffer a loss."
Sam ordered the grazing to stop. She did not want a feud to start so soon after the passing of Papa. Sam paused to think and she saw the error of encouraging her husband to make leadership decisions that he did not possess.
Bosco arrived home fuming. The bleating and the constant baa, baa of hungry sheep made him jumpy and snappy to his wife and child. He decided not to attend the evening meal. Sam protected him saying that he was unwell. He had a sleepless night. The noise of his animals was the cause. Early in the morning, he visited Victor and offered to pay to graze the sheep. Victor received the money and said,
"It is business uncle."

52

Bosco had more money than sense. He increased the size of his flock of animals that needed more food to sell later at a great profit. That was what he thought.

Victor observed Bosco saw a worried facial expression and asked,

"Uncle, you look unhappy, what's up?"

"The animals eat much more than I planned. I will buy all your bales of hay and the straw including your potatoes."

"I will sell at the going rate because I will have nothing else to sell uncle."

"No problem, I have money. When I make my profit you will envy me," said Bosco smiling.

"No, I will not envy your success but be happy because you will help me when I am in difficulties," said Victor.

Was Bosco a shrewd farmer?

At that stage of his plan, it was careless of him not to see the hurdles. His focus on profits shrouded his judgment. No one had the answer but time would tell.

Winter was near. Victor noted the winterberries in abundance. His grandad had told him that an abundance of berries was a sign of a severe winter. He had never checked whether it was fact or fiction. The conflict that was earlier a rumour became real. Bosco thought that he was still going to be very rich shortly. He paid his nephew generously.

Bosco was facing mounting problems. Lambing had surged his animal numbers. He did not have enough space to house his animals. He had no other place to obtain animal feed if Victor's supply was insufficient. He did not increase the number of farm workers to support the increased flock.

The weather however appears to be unpredictable. That was not in Bosco's expansion plan. The winter was severe and extended for longer. The lower temperatures that exceeded the average claimed the lives of many of Bosco's sheep that remained in the open enclosures. In the middle of winter, Bosco noted a large number of sheep loss. He called his nephew for advice.

Victor had empty barns and allowed the sheep to take shelter. Bosco thanked his nephew. He started to see his idea of great profits becoming a dream. He was slowly losing his jovial attitude to a sulking mood. Speaking in a tone of voice showing a lack of self-esteem, Bosco asked,

"Will you charge for this service, nephew?"

"No, it will be on mate's rates."

The good gesture demonstrated by Victor was not sufficient to elevate Bosco's mood. His bank balance was decreasing at a very fast rate, as he had to pay extra demands every day. It looked like plans that were heading to failure and he hated it. He directed his hatred at his nephew.

He walked away with heavy thoughts that Victor, his nephew, was financially stronger than he was. Victor was showing authority and Bosco had to ask for help from him, a situation he bitterly disliked. Victor was aware that his uncle felt humiliated.

The long, cold winter and the armed skirmishes of neighbouring countries disrupted many services added to the misery and took a toll on the finances of Bosco. He had overlooked that the rise in prices did not always lead to profits.

53

The final straw that affected Bosco was financial. The quality of the wool from the poorly fed animals had no value. Bosco was aware that the tumbling of the business was leading to his ruin. He could not admit that to his nephew. He was worried but too conceited to ask for help or to admit failure and did not go to see Victor every day to inspect his sheep in the barns. Victor expected Bosco to visit him.

Mama was bed-bound. Her voice was weak and meek. She indicated to Victor that she wanted to see Bosco. Victor walked to the house to find Bosco seated on his rocking chair swaying backwards and forward slowly. He was unwashed, unshaven and covered by an old blanket, showing an unmistaken gloomy face that Victor realised was a clear sign of distress.

He became disturbed to note the absence of Sam and Donald. The condition of Bosco after four days alarmed Victor.

"Uncle, are you ill?" asked Victor as he pulled a chair to sit. The weak voice of Bosco was the reason Victor had to move closer to hear as he spoke to say,

"I am ruined, broke and unable to carry on."

The bad breath and the stench of sheep muck forced Victor to slide his chair away. Victor thought that his uncle slept on the straw used for lactating sheep pens. In that state, Victor feared that Bosco was in a bad mental state and could resort to violence or self-harm.

"Uncle," Victor said in a whisper,
"Where are Sam and Donald?"
"I don't know," said Bosco.

Victor was surprised that Bosco did not know where his wife and child had gone.

"Mama wants to see you," Victor said after a moment of thought.

Bosco did not reply but tried to stand. He could not ignore Mama but could not stand. Victor held his arm attempting to give support. The thin arm of Bosco alarmed Victor, who asked,
"When did you last eat?"
"I can't remember."
Victor made him a cup of tea. In the biscuit box, he found only one and gave him saying,
"Please eat before we go to see Mama."
"Where are Sam and Donald?" Victor asked again.
"They are gone. Three days ago."
"Where did they go?"
"Donald wanted to see his grandparents."
Victor did not make sense of the long stay at his grandparents who resided just a few meters away and in plain view.
He asked again,
"What made them go and stay so long?"
Bosco was silent then he asked for some water to drink.
"Do you want some food?"
"I will have another cup of tea with Mama."
Using Victor as a prop, Bosco shuffled slowly to see Mama. Bosco entered Mama's bedroom and got out running calling Victor.
"Is that Mama in the bed?"
"Yes it is, Mama, what did you see?"
"I saw a very small lady, too small and too grey to be Mama."
Victor was not happy at Bosco's reaction. Victor went into Mama's room and informed Mama that Bosco was not well. He left gritting his teeth and said to Bosco,
"It is better to bring Sam and Donald here quickly because we have problems greater than your worries." Bosco was silent. It was not clear whether he understood what Victor meant.
Titti saw Bosco's condition and offered a glass of milk, after which Bosco said,
"Please let us go together to get them."

Sam was elated to see Bosco and Victor and emotionally said, "Thank you, Victor; I thought he did not want us. He shouted and frightened his boy. I had no choice but to leave. I thought he needed space to think." Bosco turned to Victor and said, "Sam and I had a few sharp-worded exchanges."

"Just be nice," whispered Victor. While Sam was preparing to return home, Victor sat in thought. It was heart breaking to see his uncle looking rough and dirty.

He decided to take action that would make Bosco revitalize to act normal again. His priority to heal the rift with his family appeared a success. Another challenging issue to address was the uncle's financial pressure that needed more thought. Victor had to face it if he had to take Papa's leadership position. He did not want despair to cause the stretch or split of the family bonds.

54

Victor arranged for a communal evening meal saying,
"Uncle, we have to talk to make the whole family aware of what happened and what may be about to happen."
Bosco and Sam, Victor and Titti with the children, Alena, Sofia and Donald including Harriet sat around Mama's bed, singing some cheerful songs. It was clear that the songs were appeasing as Mama indicated she needed help to sit to see her family. After a few glances, she said,
"Thank you." She resumed her sleeping position. They left the room with teary eyes.
The adults consumed their dinner in silence tolerating the occasional squabbles and jokes of the teens. Victor thought that everyone had something to say but was afraid to speak. The bored teens, intolerably noisy and the adults remaining very silent forced Victor to stand. He tapped his spoon on the glass for attention and said,
"I have noticed that uncle Bosco has been very sad and he may be suffering. He should tell us so that we share his burdens as a family and solve the problem to carry on. We will not allow any family member to suffer alone."
Sam smiled at Bosco to encourage him to talk. It was a relief to Victor when Bosco stood up to talk. He gave a detailed account of his business plans that had failed. Bosco, however, blamed everyone or everything but himself. They did not nod in approval of what he said. They looked at him silently, a sign that Bosco realised meant they were longing to hear more. He realised his talk was not convincing. He stretched his hands as a person begging and made an admission saying,

"To solve the business problems, I spent so much that I run out of money. I did not see a way out except to bow out, but Victor saved my life this morning."

Sam stood up so suddenly that the chair fell behind her. She barked,

"What? How did he save your life? Why did you not speak to me? Why did you refuse to eat and ask to be alone? Did you need space to do yourself in?"

Sam sat down without getting an answer, but her tears were flowing like raging streams.

Titti in support of Sam said,

"Why didn't you talk to your nephew, me or Mama?"

Bosco almost in tears said,

"Due to Mama's suffering, I could not disturb her. It is Victor's visit today that saved my life as I had decided tonight was the time to take action."

Bosco's answer sent a spine-tingling chill to Sam, Titti, Harriet and Victor.

No family wanted to discuss that kind of talk. Victor realised that he saw something that did not make any sense at first when he visited Bosco. He decided to say it openly. He asked everyone to sit down. The talk he was to give could be traumatic for the children. The teens were silent, as they appeared to note the serious nature of the talk. Victor stood very slowly faced Alena, Sofia and Donald and said,

"Children, please go to bed. The adults need to have a serious talk."

Alena said,

"I will take Sofia and Donald to sleep but I will stay as I am almost an adult." Alena walked to go to bed dragging her feet on the floor, a sign of discontent. She changed that attitude and walked fast when she got an emphatic refusal from her mum and dad.

Victor did not stand up but sat and was at eye level facing Bosco, Sam and Titti. He spoke in a low tone of voice, almost whispering as he said,
"This morning when I visited uncle he was in his rocking chair and his face gave me an impression of great sadness. He had not eaten for three days, he told me. He could hardly stand and when he did, I saw a rope on his chair. When he fully stood up a pen and a piece of paper fell from his lap. I did not ask him what he was going to write but now I can guess. He may have decided to do something horrid, such as to end it all."
Sam in tears said,
"Were you going to hang yourself?"
Titti asked almost immediately,
"Was the pen and paper to write your suicide note?"
Bosco nodded. There were tears with words from both Titti and Sam saying, "Why," repeating after a break between sobbing and wiping the tears away.
Victor saw the agitation, the grief and the anger. He faced Bosco and said,
"We will solve the money problem but promise not to do anything to give grief to the family." Bosco nodded and tried to smile but Victor insisted on a verbal answer. Eventually, after a slight clearing of his throat, he said,
"I will not do anything to harm the family," as he turned and embraced Sam to say sorry. They sat silently for a while probably thinking of what might have happened.
Victor's voice broke the silence when he said,
"No more talk tonight. We will never talk about this subject again, even when our tempers flare. In the morning we will solve our problems."
Sam said,
"Thank you, Victor for caring."
Titti stood up and said,

"I am going to bed now but I ask everyone to wake up and show a smile and a good mood at breakfast."
Victor stood and said "Good night all," and left.
Bosco held Sam's hand and moved to go to bed but first called out for Victor's attention and said,
"Thank you, nephew, I do not know how to repay you."
Victor said, "Sleep well, uncle."

Reminiscing in Old Age

55

Do birds ever get sad or become worried as humans do? Do they know where to get the next meal? However, they always wake up early in the morning and sing cheerfully even if it is cold, hot, wet, or dry.

Humans, however, do not show that kind of unconditional cheer. They show sorrow if they have a bad dream or have a disturbed sleep besides worrying about the food choices they have to make.

It was an early morning and no one gained a good sleep. The heinous plan of Bosco to do away with his life, the debt repayment and the deteriorating condition of Mama were matters that were not conducive to good sleep or a cheery morning.

Victor turned and twisted in bed. Titti did likewise. That was a relief, as there was no exchange of blame. Bosco revealed to Sam that he was in deep thought and did not have a chance of getting shut-eye. Sam was worried and could not sleep but did not tell Bosco.

Tired but determined to face the day and all that could probably happen, Victor walked slowly to the dining table for breakfast. He was surprised to see Bosco walking towards the breakfast table too. Before Victor could speak, he heard,

"Good morning nephew. It looks like it will be a bright day. I came to check on Mama." Victor did not know what to say but he moved quickly to greet his uncle with an embrace and a prolonged handshake. The talk and laughter that followed appeared to wake up the whole family. Titti approached Victor and Bosco with a smile and then she heard Victor saying,

"You have achieved your wish for everyone to be happy at breakfast. Bosco is in good mood and that is a good start."

All appeared happy as they sat to enjoy their mealtime. After a good rest, the youngsters were encouraged to go to do their academic work. The adults were ready for a meeting to solve problems. To keep up the mood of the relaxed theme, Sam suggested a glass or two of wine to celebrate the achievement that all family members were in a good mood. Sam was holding a tray of glass tumblers for the wine when Harriet's urgent call to go to Mama's room meant something was serious and the tray fell off her hands. All rushed into Mama's room and no one heard the noise of the breaking glass.

One by one, slowly and in silence, the family members returned to the table. They took a cushion supporting their faces as they cried profusely. They cringed hiding their faces by the cushions as they rested on the table, unable to face the world. The gathering turned into a vigil.

Victor watched Bosco, Sam, Titti and Harriet as they hid their faces by the cushions and was afraid of disturbing them. The sobbing had stopped but the body tremors could be visible. He needed help. He called Goodwood and gave him the information. Within a few minutes in the early hours of the morning, friends came and helped to make tea and anything the distraught family members could eat. Goodwood tried to calm Victor by saying,

"It is a day you would like to forget but the incident can never be forgotten. It is the good things you do today that will give you happier memories."

Victor lifted his head to look at Goodwood and said,

You are talking in riddles, what good thing can make a person happy on a day like this?"

"The visit from the neighbours, to give the family support is one of the good things," said Goodwood as he encouraged Victor and the family members to drink some fresh milk he supplied. One by one, each member of the family moved away from the table to go to freshen and talk to the well-wishers. Bosco returned after a quick wash. He, as well as the other family members, thanked the sympathizers with a shake of hands or embrace while humbly saying, "Thank you."

The family's grief was less with time. The adults gave thought to the incident and realised the meaning of what Goodwood had said.

The neighbours behaved in impeccable conduct in dealing with the passing of Mama. It was the time Victor realised the importance of appreciating the good things of that day to overcome the worrying thoughts of the sad incident. In his mind, probably in the minds of the adults of the family too, there were many thoughts in memory of his grandmother:

She became my parent and grandparent
Doubled in intensity was her love
To cherish her memory I must do
For those of the next generation

"Good morning uncle, you look good this morning," said Victor as he met Bosco on their usual farm inspection. They had to carry on as Papa had taught them and they were confident that Mama would say the same. Bosco and Victor were the eldest members of the family with seven and five decades lived respectively, they had to behave courteously as mentors. They had to tell past stories whether fact or fiction and had to ensure the youngsters viewed them as wise. It was not an easy responsibility. They had to reflect on what the mentors in the past had said to maintain the continuity of tradition.

"I laughed alone when I remember the things Mama told me that did not make sense then. However, now I can see clearly what she meant," Bosco said. Victor was about to ask Bosco a question when Sam shrieked,

"You forgot your pill."

"Do you take pills, uncle?"

"Yes, nephew, I have bilious."
Victor had heard of the condition. It was not a life-threatening illness and as his uncle could get up and go, there was no need for concern. To know more, Victor asked,
"Are you able to go round to carry on the farm inspection?"
"Oh, yes, nephew, I am not bowing out yet."
Both walked out laughing and patting each other. Victor smiled at the thought that his uncle was happy again and was forward-looking, ambitious and at times with some wit. Bosco surprised Victor when he asked,
"Do you remember when Alena said she was practically an adult? It means you are getting old, my nephew."
"Are you not getting old because your son is young?" asked Victor.
Bosco smiled and said,
"I am not getting old but I am old. When you witness seven decades in time of your life, my nephew, you start looking at the road you walked not where you are going."
"Why do you do that?" asked Victor.
Bosco shook his head and said,
"You constantly hear the words *you have arrived at your destination.*"
Victor looked sternly at Bosco and said,
"Who tells you that?" Bosco smiled and said,
"The words ring in my head day and night, but the exact destination is elusive, as I cannot pinpoint when or where."
Victor scratched his hair, which was changing to grey and said,
"Uncle, you are not travelling you are always at home."
Bosco said,
"Life is like travelling but the destination is my home, the end of all the travelling after seven decades."
Victor smiled as Bosco was showing the wisdom of Papa and Mama and he did not know how to reply. He heard Bosco saying,
"In the early days business occupied my mind. I did not see how quickly the children were close to becoming adults. I better talk

more to Donald and you should not delay talking to Alena and Sofia."

Victor was proud to hear Bosco speaking wisely.

Bosco was happy to share his past with his nephew. It was going to be a long story, Victor thought as Bosco moved quickly towards a mound where they usually sit to exchange stories of the past.

56

Bosco put his long coat on the mound as a cushion to sit on. He had in the past a painful bottom from sitting for a long time.

"Nephew, I am rushing to tell the story of my life before my sunset occurs," said Bosco.

Bosco thought that it was a suitable spot to sit for telling a past story while enjoying the spans of their farm and imagining a future without his presence. That preparation made Victor eager to hear about his uncle's past. He did not have long to wait.

"It is a long story, my nephew. I hope you will be a good patient listener. It started after finishing basic education. I did not pass. I failed as many expected. I was a bully and full of pride that I was the greatest and richest. I saw studying as an irrelevant activity. I always had pocket money for sweets. I had never known the feeling of hunger and I did not expect to have a problem getting food. My adoptive parents sent me to expensive training schools, which I attended for only a few weeks and left. A short time at a technical college to fix machinery and a brief moment to learn drama failed.

Finally, my parents tired of my complaints asked me to tell them what I wanted to do. When I said, I wanted to go to Moshiana. My father asked what I expected to do there and I said to get a city job. I thought Moshiana was the big city as those I saw in magazines. They let me go. I think they knew how bad Moshiana was, but they thought that the town would teach me humility and that could be a lesson for my insolence. They gave me a good sum of money to spend with the condition that it was all they would give until I returned. I left without looking back. I thought I was free from the school learning that I did not like. In the town, I went to eat in expensive hotels and made friends with government officers who wore suits. I wore a suit too. My money was running out fast. The owners of the apartment threw me out. My suit was worn and torn. I moved to cheaper areas. I had to find work and that was when I met Harriet's father. I did not know anything about him but he gave me a job to drive workers to various sites.

In the cheaper housing areas, I was mixing with beggars and criminals. I was living rough in a room with nothing but a bed. During those times, I could afford basic things but never a luxury. I met Timson. He was my companion when I had the last penny. I bought him a pint and he made me cheerful by telling me town stories. He was a beggar and took an opportunity of work whenever he chanced. I reached the lowest stage in humanity when I had no food and at times did not know when I could eat until my next payday. I joined a bunch of criminals, they said just one job and we could be rich. Who would not take such a chance when you are close to becoming a beggar? They promised riches but they were very poor. I must have been a fool to believe them.

"You were not as wise as you are now, uncle," Victor interrupted the story.

The robbery failed and Timson saved me from the police. I decided to return home.

I had no money for fare to go home. I hitched a ride in a truck that was returning from carrying pigs to the town.

It was an overnight ride in cages that housed pigs. They had not cleaned the pens. There were the leftover fruits that the pigs ate. You can imagine, nephew, I ate those fruits as I was starving. The truck dropped me some miles from home when I had to walk. I was hungry, dirty and smelling like a pigsty. Above all, I was angry with everyone and everything. When I reached home, the most shameful thing was that I saw my mother crying because of the smell, the filth and the malnourished state of health I showed. I just stood there saying nothing. What could I say? Could I tell how I rummaged in the feeding bowls of pigs to

see if any edible fruits were left, as I was very hungry? Could I say how I spent the money they gave as if there was no tomorrow? I was then a man and a disappointment to my rich parents. They allowed me into my room to wash and change. The clothes I had left at home were buggy. I had to use a belt to hold up my trousers. I looked in the mirror, my teeth were no longer white and in my black hair were some streaks of grey hair. When I greeted my parents, my father had no words but nodded but mum cried. To ease the tension I asked to go to check on my old friends. I returned home sad, I did not find any of my school friends. Their family told me that they were in faraway places in pursuit of advancements. That information made me look like a total failure. I returned home devoid of the will to live. I thought the world had discarded me as a spent cartridge. At the dinner table, I was silent, unable to speak and I faced down at the plate of food that I was unable to eat although I was hungry. My adoptive dad knew I had failed and politely asked,
"What do you want to do? I do not think you are happy."
I looked up at him; I nearly cried but put on a brave face and said,
"I want a farm, I hate cities."
My mum started crying aloud and then said,
"You know nothing about farms, crops or animals."
My father said,
"Let us give him what he wants."
The next day they travelled with me to this place and bought one small, neglected farm. The sheep and the pigs were included. They paid and I was a farmer, but probably the most stupid one.
"You got it easy uncle, Papa and Mama and my dad worked so hard to acquire their farm and that is why they did not want to sell," said Victor.
"In the early days, I did not understand the pain of others, I thought of my own needs, without the care to others," said Bosco.

"Have you been happy as a farmer?" asked Victor.

"The truth nephew is that I did not know what to do. I got ideas from any who wanted to be my friend. Many became friendly to enjoy the drinks I generously offered. When you threatened to buy my farm, you could succeed because for years I was losing cash. I was afraid but had to put on a brave face. I could not admit it. My parents were dead. Where could I go for help? I was about to lose the will to live when you discovered me and saved my farm and possibly my life. In the recent incident, you proved to be my rescuer. You have no idea how important you are to me nephew."

Victor patted Bosco's shoulder and said,

"You are my uncle and the only one I know or will ever know."

It was past time for lunch; they stood, stretched their muscles and patted their painful bottoms as they walked home, leaving the storytelling for another day.

57

The evening started like any other.

The sun setting as usual,
The birds chirp saying goodnight,
Many animals walk to their dens,
Some nocturnal ones prepare to come out
Finally, the earth turns to extinguish the sunlight

The family was expecting to view the full moon and were not disappointed. Victor noticed that Bosco was smiling more than usual as he watched the bright moon. He guessed that something was in his mind. A glass of wine after an evening meal appeared to have made Bosco relaxed and eager to talk. Victor thought that Bosco had another life-lived story that he wanted to tell the family.
"You look to be happy uncle, what's up?"
"Nephew, did you know it is my birthday today?"
"No uncle, you never mentioned that before and we never asked."
"I did not mention it because the thought of the birthday and the separation from my mother were bitter memories. Now knowing my mother's circumstances at that time, as disclosed by my uncle, Timson, I have accepted to celebrate it. I must say thank you for your research. At seventy-five, I am celebrating my birthday for the first time. Shall I call it my first birthday or my seventy fifth one? I can now make peace with my deceased mother and feel happy."
"It was your kindness to a beggar that resulted in discovering your uncle. The uncle helped you to reconcile your thoughts with your mother. Now you are a happy old man," said Victor.

"It started with my failure to live in Moshiana. That failure, good or bad led to a chain of events that led to my returning to Moshiana. Meeting Timson again led to my finding him to be my uncle," said Bosco.

Titti heard that chat of uncle and nephew and then said,

"It would be impossible to find you if you had succeeded in purchasing this farm."

"If this farm did not exist, after college, I would have gone to work in the city, visit Papa only occasionally and the need to search would not arise," said Victor.

Titti said,

"The good thing was the failure to purchase the farm."

Victor said,

"I have never thought a failure to be a good thing."

"It was a good thing indeed. Now I can see the benefits failure can bring," said Bosco.

"Your vitriolic conflict with Papa, ugly as it was, must have a beneficial role in finding you because it was when I wished you were dead that I remembered my dead parents and became curious to know about them," said Victor.

"Do you think you benefited from your acts of wickedness?" asked Titti.

"My insecurity, brought about by my inferiority complex caused by many previous failures had a role in making me bitter, rude and wicked," said Bosco.

"You are happy now, with no worries, your failures are now insignificant," said Victor.

Titti said,

"Did I hear it was a birthday?"

Bosco smiled and said,

"It is my first one in seven decades."

Titti said,

"It is late now but we must celebrate this unique birthday. The youngsters will sing a song to make it an indelible moment for our beloved uncle. Tonight, however, there is reason to make it a special festive. I will be back soon," said Titti smiling as she left.

Bosco faced his nephew and said,

"It is amazing how in old age you view previous failures as part of the incidents that brought happiness or success. I can now enjoy the shining moon and the bright stars bringing calm and tranquillity to all creatures including me. Before I was an uncle, I never enjoyed this vision of nature."

Titti returned with Alena, Sofia, Donald, Sam and Harriet and said,

"We are going to celebrate your birthday under the bright moonlit night."

Bosco smiled and said,

"I will not find proper words to thank you. I saw for seven decades, many moonlit nights but I was never calm or relaxed when I was just Bosco. Nothing appeared to me as good until Victor discovered me as an uncle."

"In the past uncle, you suffered despite the money from your rich parents. You needed love. When you discovered real love, the world was beautiful and you appreciated it. Love liberated you although it happened in your old age. I hope you are no longer bitter or feel unloved," said Victor.

"I wonder how I survived with that bitterness within me all that time. I was healthy but miserable, now I have health concerns but I am happy," said Bosco.

It was late in the night but the moon was bright. Keen to hear about Bosco's health issues, Titti opened another bottle of wine. After a sip, Bosco wasted no time and said,

"Ageing is not always synonymous with decline, worsening or deterioration. Animate or non-animate creations suffer the same fate of deterioration with time. The body systems work tirelessly for many decades to upkeep the body constantly rejuvenated.

There are small undesirable changes but nothing to fear. The airways, for example, sustain life through the exchange of air between the body and the atmosphere. Unwanted materials in the air may enter the body and clog the air passages restraining airflow. After seven decades, signs of inefficiency such as laboured breathing will manifest. Other problems sneak slowly one after another and after a while, there is a long list of problems."
"You do not appear to have such problem," said Titti.
"My body systems are fighting the deterioration. There is success but not total," said Bosco.

The moon had moved significantly in the sky, the wine bottle was empty, the eyes were at times closing and the chat ended but to continue another day.
In bed, full of thoughts and unable to fall asleep, Victor reflected on the chat with his uncle:

> *Old age is not always equal to a lack of wellness*
> *The zest to cling to sweet life is endless*
> *There is fear of the end, becoming lifeless*
> *Happy with his family uncle is fearless*

Everyone was in silence thinking or anticipating the next story from Bosco. Victor thought that family stories could do wonders for the wellness of a family. He looked at Bosco hoping he would continue with his many stories but saw something he had not noticed before. It was a scar under his left ear. He asked,
"Uncle, how did you get that scar?"
"In old age, scars remind you of your youth activities," said Bosco smiling and touching the scar.

58

Bosco woke up early in the morning showing great cheer as he greeted his nephew saying,
"I could not sleep."
"Why?" asked Victor.
"You asked me a question *'how old are you now?* I rehearsed the birthday song I heard last night, seventy-five times to give the answer," said Bosco laughing hilariously.
The cheerful look of Bosco persisted the whole day and everyone was looking forward to hearing the story about the scar below the left ear.
"When I was young with more money than sense, I was in a bar drinking where the young and rich office workers met. It was a place to see those you thought were in your league. I was neither educated nor working in an office but I had cash and wore a suit and that was the password. The town scroungers knew the rich young never finished the drinks and they frequented the bar for leftover drinks. They helped to collect empty beer glasses and at times helped to wash up to get a free drink and a warm place to stay. One beggar lifted a pint by mistake thinking someone had abandoned it. The pint of beer belonged to a person who had gone for a call of nature. I sat at the next table alone, minding my own business and listening to the arrogant chat of the rich young office workers. I could hear their conceited talk and see their new expensive shirts. They referred to those who beg as rats and vermin that live in the sewers. On returning to find the beggar with his pint, he lost his temper. I offered to buy him a pint to stop the rage. He got angrier. One rich young man wearing a thick gold chain over his shirt said,
"Slap the beggar and make him vamoose."
I stood up and said,

"You cannot slap an old man. I will defend him."
Had I suddenly changed from a bully to a defender?
Two beggars were standing close to him, intending to share that pint. They faced me and said,
"Thank you, you are a good man."
No one had described me like that before. I became empowered. I remember my days as a bully and the fear I gave towards those I bullied. I told the angry man,
"Go and get another pint if you can afford it."
The beggars in unison said in a smirk,
"Get another pint. You can afford can't you?"
He was livid, he was biting his lower lip and then suddenly he took his hand out of his pocket and rapidly swung it to pass under my ear. There was a sharp pain and I staggered to fall but a very dirty beggar standing behind me prevented me from falling. I panicked when he shouted,
"He is bleeding." I started to breathe fast. I was alone in that town. If I died, they would lay my body to rest in a shallow grave in the Moshiana cemetery. Eventually, you could not read the headstones covered in soot. It was a thought that made me very frightened. The fact that I was bleeding and the great fear was that I was going to die.
The beggar who took the pint smashed the glass on the angry man's head. He fell. The other drinkers noted the skirmishes at our corner and approached the table to investigate. Those without smart suits started to attack those smartly dressed. Their smartly dressed women friends started to run away. Some beggars used the commotion to manhandle some of the well-dressed women. The drinking establishment became completely riotous. Drinks and glasses thrown randomly into the air were a potential danger to everyone. It was like a mob scene where chairs, bottles, glass tumblers and ashtrays became weapons. The scrounger who prevented me from falling took me outside for safety and tied my neck with his off-white handkerchief. I

thanked him and left very quickly. Luckily, the scratch did not need medical attention. I started to dislike the wealthy and respected those who lived on the margins like the rough sleepers. When my money ran out, I needed company and the beggars did not judge me. That path of life led me to Timson, the beggar I came to know as my beloved uncle. The act of buying him a drink was my way of trying to do a good turn to the beggar that had helped me."

Victor was proud that his uncle was showing wisdom, as he was advancing in years graciously and thought,

> *Wisdom in ageing is an honour*
> *Freely given not from scholarly power*
> *As a gift, it is only for a few*
> *Glad that my uncle is one of the few*

"In growing old gracefully, do you still have ambitions to do what you could not do in your youth?" Victor asked to break the silence that had persisted longer than socially acceptable after the story.

"Yes, nephew, I do have lots of wishes but the ability to get up and go has gone and what is left is accepting with glee the process of growing old gracefully."

"What makes you enjoy life and keep you happy on daily basis?" asked Alena.

Bosco brushed his long beard. It was a sign of happiness. He picked up his wine glass, took a sip and then said,

"When I see you Alena, Sofia and Donald growing up energetically as you go and come from your daily tasks of study and work, I feel proud. It is an achievement of continuity of the cycle of life that Papa and Mama helped to make it as perfect as possible."

59

Victor was thinking of the future, and Uncle Bosco was thinking of the past. The act of looking at the past has the benefit of generating at times hilarious stories from tricky situations.
The future, however, is not a story but a life of hope, which helps to overcome despair.

Victor looked in the mirror one early morning. All his hair had changed from black to nearly all greyish white. He called Titti and said,
"I am growing old gracefully but I am afraid."
"What do you fear?" asked Titti.
"The previously efficient systems appear to fail."
"What do you mean?" asked Titti.
"There are body functions that are not fully in my control and are likely to cause embarrassment making attending social functions undesirable."
"Do not panic that happens to all of us," said Titti trying to calm him.

One day after harvest season, Bosco was happily reciting good moments of the past while Victor was hoping and wishing for the best future that life could provide. The exchange came to an abrupt halt when Victor resting on a soft settee and Bosco, from his sitting position on his rocking chair at the veranda, saw a shiny silver-painted car approaching the house. It was an expensive car; a chauffeur dressed in an immaculately smart uniform came out and opened the rear door. A tall slim man, dressed in an expensive suit slowly came out. The arrival of the car attracted the attention of Victor, Titti, Sam, Harriet, Alena, Sofia and Donald who came to the veranda to admire it. Victor,

Titti, Sam and Harriet watched the young slim man in a suit, trying to guess his intentions for the visit. The man looked around as if he were an inspector of some sort. The long-time taken in looking around enraged Victor. He thought it was a land buyer. That bitter thought made Victor start to grind his teeth. Titti quickly noted her husband's reaction. She averted a confrontation by moving close to the car, and said to the tall thin man,
"What can we do for you, Sir?"
"I am not a Sir, ma'am but an ordinary man who works hard enjoying the fruits of my work and a little extra of what the world can offer." Victor relaxed but Bosco noticed something that made him stand up. Titti turned to face Victor, not to seek help in questioning but to see whether the voice was like one they had heard before. Victor wanted to know the purpose of the visit above all. He stepped forward faced the man who appeared much taller at close range and said,
"What is the purpose of your visit?"
"I am a businessman from the city. I buy livestock and sell it to city butchers. I am looking for anyone who can give the level of business needed. Sheep farms that produce wool may benefit from this business venture too. I am responding to an advert by your local bank that needs the opening of business accounts. I sell farm machinery for cash or on lease to farmers. I am a friendly businessman."
Harriet whispered to Titti,
"I was the one who sent out those adverts. I hope I will get a promotion."
Victor saw opportunities; he boldly moved close and asked,
"What is your name?"
"Robby is my name and what is yours?"
"Victor and the gentleman behind me is my uncle, Bosco, this is my wife Titti, our Cousin Harriet and the children Donald, Alena and Sofia."

"I am happy to know you and your family," said Robby.
"We can supply sheep, pigs and wool if you fulfil our conditions." Robby turned to look at the car. He did not want Victor to see the smile revealing his excitement. Bosco was excited too. He whispered to Victor,
"Let us show him the farm."
"Not yet, uncle, we have to get the upper hand." Victor turned and saw Robby and the chauffeur whispering. He suspected they too were planning something when Titti whispered to Victor and said,
"Try to make him smile. There is something about him but I cannot put my finger on it." Victor remembered Papa's tactics and coughed to get Robby's attention by saying,
"We cannot do business without knowing a little of your background"
Robby stood boldly in front of his audience and said,
"My name is Robby Tim McNaff. I am the boss of City Suppliers." Victor had to think fast and said,
"I hope your company has got accreditation we could check. You can, however, assure me there will not be any need to investigate."
Robby looked at Victor and saw intellect. He smiled to give him confidence and said,
"You will be satisfied. Check with the bank. Now let me know who is selling and who will buy my machinery."
The discussion between Robby and Victor went on for a while about the machinery. That gave time for Titti and Bosco to exchange some observations. Titti and Bosco were sure that Robby has some family lineage based on his voice, the teeth and the eyes that resemble those of the late Uncle Timson.
Bosco said,
"I will question him after the business agreement."
Titti said,

"Not the kinds of questioning you do uncle. The situation is delicate. He is rich. He thinks we want some of the wealth by claiming family connection."
Victor asked Robby for a moment to get a green light on the price before the agreement. Victor whispered to Titti, Sam and Bosco and then he said to Robby,
"What price per kilogram are you prepared to pay?"
Robby approached the chauffeur who handed him a calculator. He did some calculations and showed a figure to Victor and said,
"I hope that is satisfactory."
Victor took the calculator and showed the figure to Bosco who whispered,
"It is far better than any previous offers. We can supply monthly or twice per month." Victor shielded his uncle to avoid revealing a quick acceptance. Victor held Bosco's hand firmly and said,
"Let me ask a little bit more."
Bosco replied quickly,
"Go on nephew you are doing well."
Victor approached Robby holding the calculator and said,
"We want you to shine among your customers. A little increase of ten units will enable us to feed your animals with special exotic recipes, a guarded family secret passed through the generations. The unique taste of the meat will make you famous." Robby looked at Victor, took his calculator, pressed some keys that made very fast tic-tac-tic-tac noises, paused, smiled, showed to Victor and then said,
"See this value."
Victor took the calculator again and showed his uncle the new higher figure without a word.
Bosco said,
"You have made us rich, nephew, Harriet will facilitate our transactions."
Victor handed back the calculator to Robby and nodded to show agreement followed by a handshake that sealed the deal.

Before Victor could say a word, Titti had to get his attention by pulling his ear and whispering.

"Before Robby goes away we must know. We cannot let him go before we know."

"What is this you must know now?" Victor asked angrily.

Bosco tapped Victor's shoulder and whispered,

"Nephew, Titti and I have a hunch about Robby."

"What kind of hunch?" Victor asked whispering.

Titti close to Victor's ear in a firm voice said,

"Just check, you may have a cousin. Do not let the opportunity go to waste." Victor smiled and said to Titti loudly,

"We must try to get the best value for our hard work."

Victor noted that Robby appeared worried from the numerous whispering and said,

"We try this price for the first six months. I promise after that you will make a bigger offer." Robby looked at Victor, smiled and then said,

"We will be happy to try."

Victor nodded, smiled and after a brief pause said,

"Thank you for the business."

Robby said,

"Thank you and I am glad to do business with you."

Victor and Robby extended their hands for a shake to seal the business agreement. While they were holding hands, Victor said to Robby,

"Before you go, there is a small matter of observation I would like to talk to you about. I hope you will not mind."

Victor was surprised when Robby said,

"There is a small matter I would like to ask you too."

Victor said,

"You look and sound like someone I have met before. Now we are business partners, we can indulge in some pleasantries."

A Dot on the Timeline

Robby told the chauffeur to move the car to a shade. It was just to get to speak to Victor in privacy. He moved closer to Victor who held his hand and led him to the veranda.

Robby did not only surprise Victor but created a commotion by saying,

"The old man, you introduced as your uncle, rocking on that chair looked like a picture of a man in a newspaper who sold coal in Moshiana. My mum called him Tim."

60

Bosco, Victor and Titti became very attentive. Their eyes opened wide. If their ears could swirl like those of a fox, they would point forward towards Robby so as not to miss any said word. Robby instinctively suspected that the reference to the look of Bosco or the name Tim hit a nerve. Robby saw the reaction but waited for a verbal response.

"Tell us more about Tim," The interest in Tim made Robby reveal more by saying,

"Tim was my father. I have not met him but I have searched for him in Moshiana and failed. Many in Moshiana knew a man called Timson who sold coal and when they saw the picture, they agreed it looked like him. Tim and my mum met once. It was a short liaison and mum said he never knew he left her with a child. He was kind; he gave my mum a ring, maybe to say thank you for that favour and my mum gave him a neckless with a cross. My mother gave me the ring. She cut off all links with him when she met the man who became my stepdad. That is all about me. Now, let me hear what made you keen to know more."

All the adults sat still and in silence. Robby became agitated at the depth of the silence and said,

"Have I said something annoying?"

Titti said,

"Would you want a glass of wine? We have a small brewery here."

"I will enjoy it but another day," said Robby.

Victor stood up and said,

"We should arrange another meeting soon. Please bring the picture and the ring and we will talk more."

Robby did not expect that type of reception to his background story. Born out of wedlock was no longer a stigma. Many thoughts weighed on his mind. He was trying to make sense of the situation. He was annoyed at the great silence; he walked fast to his car and drove off.

The chauffeur saw the gloomy face of Robby and asked, "Boss, what is the matter?"

"Just drive, I am thinking."

Victor, Titti and Bosco assessed the situation.

Titti said,

"I think he left because of the interest in the ring and the picture."

"He was shocked by our silence. He will return because he thinks we were hiding something. He may have more questions for Bosco," said Victor.

"He probably thinks I am Tim. That thought will bring him back very soon," said Bosco.

In his bed, Victor thought of the activities that happened earlier. He was proud of the business prospects and hoped that the family would become financially secure. About Robby, in the event he was Timson's son, both Bosco and Victor would gain a cousin, but he did not care whether it was the first second or third relation of cousins. He knew that everyone would be happy, as Robby's story was likely to be true. He smiled knowing that a rich man like Robby would not want a poor family of farmers unless there was some kind of bond.

You will be disturbed to awaken to the loud sounds of heavy machinery early in the morning. You will become agitated if the sounds appear to approach your residence without anyone first informing you. That was bound to make anyone wake up with questions that would indicate anger. This was what happened early the morning after the visit of Robby.

The noise of tractors, trucks and small cars reversing and parking in front of Papa's house made the family wake up. Victor and Bosco went and stood on the veranda waiting for an explanation. Lined behind them were the other members of the family

looking on. A big man moved closer to Victor and Bosco, handed them a big envelope and said,
"The boss wants a hundred sheep now and here is the order and payment." Bosco handed the envelope to Victor to open. Victor saw the order list and the cash.
Sam and the farm workers filled the first order at speed and in cheerfulness that deserved admiration.
Bosco did not count the cash; he signed the chit and thanked them as they left. He was sure it was not a small sum. It was sufficient to cover the sale and there was more. Victor and Bosco thought that there was a reason for the generous offer and payment.
Robby arrived later in the afternoon to meet with smiling faces. He accepted a glass of wine and sat looking at Bosco and said,
"You may or may not be my father but you look very much alike. If you claim to be the one, I will accept you without question."
Robby produced the newspaper picture of the man who sold coal in Moshiana. Bosco looked at it, he gave it to Victor and Titti but when Alena saw it, she shouted,
"That is uncle Timson's picture."
"Are you sure?" asked Robby.
"Yes, he used to live here but died."
Robby stood up, bent forward towards the face of the little girl and asked,
"Are you sure?"
Victor asked to see the ring. There was an imprint of 'S' on it. Bosco took a chain with a cross that Uncle Timson used to wear and showed it to Robby saying,
"If your mother can remember the chain and confirm it as the chain she gave to Tim then we have a serious issue to discuss together." Robby had not seen that chain, he had heard about it and he was keen to see it. He looked at it so intensely that the others thought he saw something. He called the chauffeur who took the chain and returned to the car. She opened the rear seat

door and a woman came out, looked at the chain, and quickly walked to meet Bosco and Victor. She placed her thumb next to an imprint of 'B' on the cross and asked,
"Where and how did you get this?"
"What does 'B' stand for?" asked Victor.
"Barbara, my name but he gave me a ring with an 'S' imprint. I accepted at that time but as his name was Tim, I was not impressed. I thought the ring belonged to another woman with a name that started with 'S'."
Bosco said,
"Please, have a drink of our wine, we may have the explanation." Bosco was proud to tell Robby's mother that the ring given by Timson to her was likely to have belonged to Sofia, his sister who died. A confirmation from Robby's mum proved the chain to be that she gave to Tim.
A raucous talk of excitement, that overlooked the news that Timson was no longer alive, went on for hours. The chat was not without emotions. Titti, Victor, Bosco and Sam took turns telling Robby about his father and the family circle. It was time to show Robby where his father was resting in peace. Victor walked slowly at the front as the rest walked and followed in single file a winding path that resembled a snake sand path. They walked through the woodland part of their farm left to grow naturally but intended for future development. When they arrived at a mound that was fenced by trees, Victor stopped and the whole family stopped. He beckoned Robby to come closer. He whispered,
"This is where our departed family will reside for eternity. Come and talk to your father."
Robby walked forward into the enclosure and saw the gravestone. He read, *'Timson, an uncle and Great uncle.'*
Robby asked,
"Can I add that he was a dad?"
Bosco and Victor without hesitation and in unison said,

"Yes."
Robby had so far behaved assertively but to know he was placing a flower on his father's grave, he could not avoid shedding a tear. Bosco and Victor embraced him, and all three men appeared to have tears flowing. It was a dot on the time of life of Robby and a spot on the family timeline. Robby was proud that he was sitting with his family.

He stood up and announced,

"My three children will visit this family soon. I will want them to work here for a while to learn how the food they eat grows and how hard it is to get the steak they demand to eat every weekend."

He made all the family laugh when he said,

"I would like to return after two weeks to find them with the knowledge of feeding pigs, cleaning their pens and if there is milking to do; I would like them to get that experience. Tell them there is no pocket money for failures."

Titti asked,

"Why are you harsh like that?"

"They should experience the muck their grandad endured and will grow to respect those who supply us with food."

Victor and Bosco remembered how successful it was when they involved their children in the farm work and they replied,

"We will make sure they pass."

Victor looked at Titti and said,

"Papa and Mama's wish for a growing family is achieved, and my hope for a future of happiness is granted. I think Bosco will see this family addition as happiness that facilitates his wish of growing old gracefully and give hope for a similar future for all of us that will follow."

Acknowledgements

To my wife:
Dora, thank you for your patience and your strength for all you do for us. You endure long periods of silence when I write. I appreciate the help in the choice of words
you contribute to my writing.

To our children:
The struggles you show to improve your lives are heart-warming to your parents and we support you.

To the grandchildren:
The love you show to us is immeasurable and make sure you surpass your parents to make their happiness perfect.

To the great-grandchildren:
You are now young but we are already aware of your great potential. You will make your parents proud.
The newest greatgrandchildren born in 2024 are:
Sanayah Laila Lidiu (DOB-03/02/2024)
Anas Junaide Bowers (DOB 14/06/2024)
You will have so much love. Make your parents proud as you grow into adulthood.

Last but not least:
I say thanks to all members of **Moseley Writers' Group** *for the critique and encouragement.*

About Paul V. Mroso

Dr Paul V. Mroso was born near the slopes of Mount Kilimanjaro, obtaining his basic education in Tanzania. He completed his BSc and PhD in Pharmaceutical Sciences in the UK. Previous experience was as a manager in Keko Pharmaceutical plant in Dar es Salaam.

He worked until retirement as a Community Pharmacist in England.

Paul spends his free time in retirement as a gardener, teaching his grandchildren the art and task of growing and enjoying self-grown products. The grandchildren have the pleasure of enjoying the jam made from fruits grown in the garden.

Volunteering and writing, including solving some Sudoku challenges, with some periods for holidays, tend to take all his time.

Other Books by Paul V. Mroso

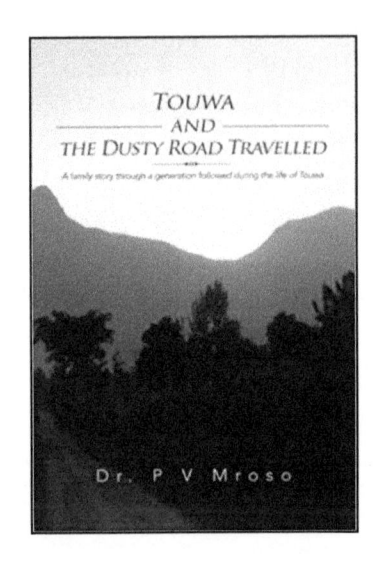

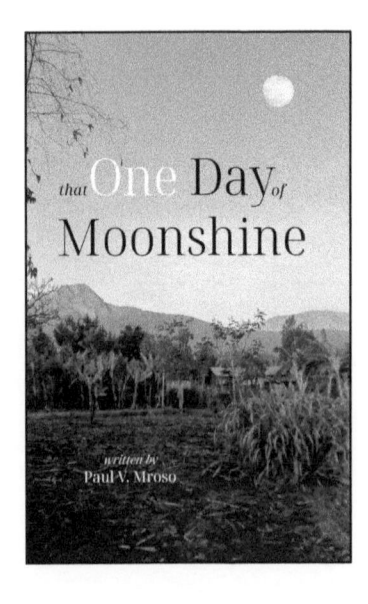

*Available worldwide from Amazon
and all good Bookstores*

Michael Terence
Publishing

www.mtp.agency

mtp.agency

@mtp_agency

Milton Keynes UK
Ingram Content Group UK Ltd.
UKHW030903011224
451693UK00001B/124